The Simple Book of Not-So-Simple Puzzles

The Simple Book of Not-So-Simple Puzzles

Serhiy Grabarchuk
Peter Grabarchuk
Serhiy Grabarchuk, Jr.

A K Peters
Wellesley, Massachusetts

Editorial, Sales, and Customer Service Office
A K Peters, Ltd.
888 Worcester Street, Suite 230
Wellesley, MA 02482
www.akpeters.com

Library of Congress Cataloging-in-Publication Data

Grabarchuk, Serhiy.
The simple book of not-so-simple puzzles / Serhiy Grabarchuk, Peter Grabarchuk Serhiy Grabarchuk, Jr.
p. cm.
ISBN 978-1-56881-418-6 (alk. paper)
1. Puzzles. I. Grabarchuk, Peter. II. Grabarchuk, Serhiy, Jr. III. Title.
GV1493.G73 2008
793.73--dc22

2007049397

Printed in the United States of America

12 11 10 09 08 10 9 8 7 6 5 4 3 2 1

Contents

Preface

This book derives from *Puzzle Miniatures*, Volumes 1–3, three small books published by Serhiy Grabarchuk Puzzles in 1998, 2003, and 2005, respectively. We call "puzzle miniatures" tricky challenges that are simple looking, but have not-so-simple solutions.

The book presents more than 100 such "mini-puzzles" of different kinds: assembling, mathematical, logical, visual, spatial, number, word, dissection, dividing, dot-connecting, matchstick, coin, and some other kinds of challenges and brainteasers.

To learn more about the book and its puzzles, you can visit our respective websites: Serhiy Grabarchuk (www.ageofpuzzles.com), Peter Grabarchuk (www.peterpuzzle.com), and Serhiy Grabarchuk, Jr. (www.unipuzzle.com).

While creating and designing the puzzles for this book, we received a great amount of support from the other members of our big puzzle family, which includes also Tanya Grabarchuk, Helen Homa, Kate Grabarchuk, and our Ma and Granny, Galina. All puzzles were selected to make puzzling with them as interesting and pleasant as possible. Our hope is that the puzzles gathered in this collection will bring many fun moments to everybody who tackles them. Happy puzzling!

Puzzles

Introduction to Puzzles

Puzzles in this collection for the most part are presented in quite a straightforward and clear way and include detailed drawings and descriptions. For some types of puzzles there are certain general rules that should be observed, unless otherwise stated.

Difficulty Levels

Puzzles in this collection have a wide range of difficulty levels and include a mix of easy, moderate, and difficult challenges. There are no reliable criteria to determine difficulty levels of puzzles precisely and universally. Moreover, it depends greatly on the solver's skills and preferences which puzzle will turn out to be easy, moderate, or hard. So, puzzles of different difficulty levels are distributed evenly throughout the book based on our evaluations.

Grids and Patterns

Many puzzles in the book have illustrations with grids and/or patterns within (or around) puzzle shapes in order to show exact proportions for your convenience.

Dividing Puzzles

These are challenges in which you have to dissect or divide some shapes with lines or with the help of matchsticks, elastic bands, or different transparent frames.

When you are asked to divide a shape into some number of parts of the *same area*, this means that the areas of the parts are equal, although their shapes are not necessarily the same. Remember that every shape's area must be fully used when you solve the puzzle. In dissection puzzles where the shapes are divided and then rearranged into other shapes, pieces can be rotated but not turned over or overlapped.

When you are asked to divide a shape into some number of *congruent* pieces, the outlines of the resulting pieces must be exactly the same. Congruent pieces may be mirror images of each other.

If you are asked to find *different* pieces, this means that no two pieces can have the same size *and* shape. (Note that congruent pieces are not different.)

Matchstick Puzzles

These puzzles use matchsticks, but you can use toothpicks, pencils, or any other stick-like things of equal length. Matchsticks in all such puzzles are line segments of length 1, unless another length is specified. Generally, matchsticks must be placed so that each end touches another matchstick. There are some exceptions—puzzles in which the matchsticks are allowed to have *loose ends,* ends that do *not* touch another matchstick. Such exceptions are clearly noted. Matchsticks must not overlap each other, unless it is stated in the instructions. You are not allowed to break or bend matchsticks.

Coin Puzzles

Lighter and darker circles always represent different sides of coins; lighter circles are "heads-up" coins, while darker ones are "tails-up" coins. You can use other circular pieces instead of coins as long as you can identify the head and tail.

For *sliding-coin* puzzles, the coins should slide over the table or within a board without lifting or jumping over other coins, unless otherwise stated. In most coin puzzles a move is legal only if the moved coin touches at least two other coins in its final position. This "double-touch" rule is required since it ensures an exact position of the moved coin(s) in its new place.

Three types of coin moves will be performed in coin puzzles in this book: single-coin moves, pair-coin (including "horizontal-pair") moves, and trio-coin moves.

- To perform a *single-coin move,* you move one coin at a time so that in its new position it touches at least two other coins from the unmoved group, except the case when this coin (marked as @) makes a straight line triad, like O@O.

- To perform a *pair-coin move*, choose any two adjacent coins, and move them orthogonally (up or down, left or right, without rotation) so that in the new position either both moved coins touch at least one unmoved coin or one of the moved coins touches two unmoved coins.
- To perform a *trio-coin move*, choose any three adjacent coins that form a small equilateral triangle, and move them orthogonally so that in the new position either at least two of the moved coins touch at least one unmoved coin or one of the moved coins touches two unmoved coins.

When start and goal positions for coin puzzles are shown in special double diagrams, their shapes and patterns must look and be oriented exactly as shown in the corresponding diagrams, unless otherwise allowed in the instructions to a particular puzzle. Note that the goal position will not necessarily be in the same place as the start position was.

Counting Puzzles

In these challenges you will be asked to count all possible shapes hidden in some structure or find the maximal number of figures that can be assembled from some sets of shapes or contours. All such puzzles will require very careful observation and systematic counting. The trickiest of these puzzles are challenges with differing or nonequal distances. They are presented with special boards and push-pins that you should place in the boards' holes. As far as we know, the first puzzle of this type was described by Japanese puzzle guru Nob Yoshigahara.

Quadro Block

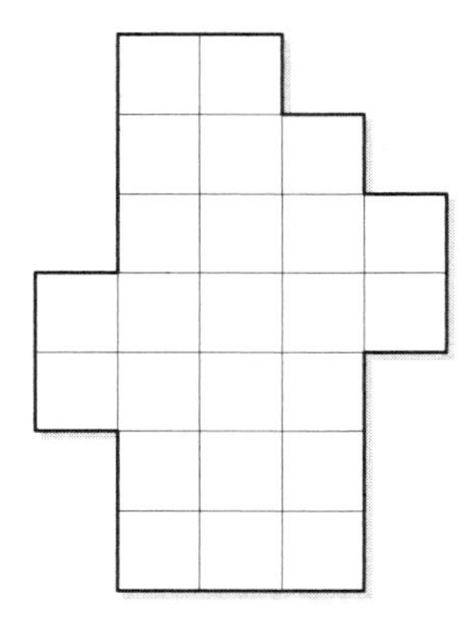

Divide this block into four congruent parts.

Solution on page 64.

Five Buttons

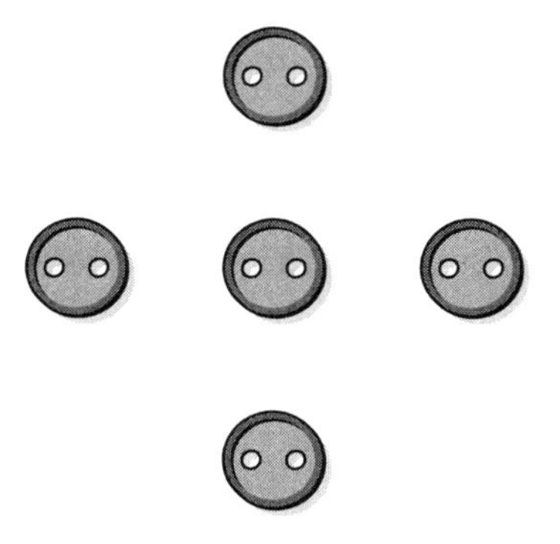

Connect the buttons with four line segments linked at the ends to form a closed route, passing through the centers of the buttons and visiting each button just once. The lines cannot cross one another.

Solution on page 64.

Twenty-Four-Seven

A big billboard off the road features the "24/7/365" sequence on it. Obviously, it should mean the business never closes, but considering it as a true logical sequence, something is wrong with it. Can you say what?

Solution on page 64.

Out of the Y

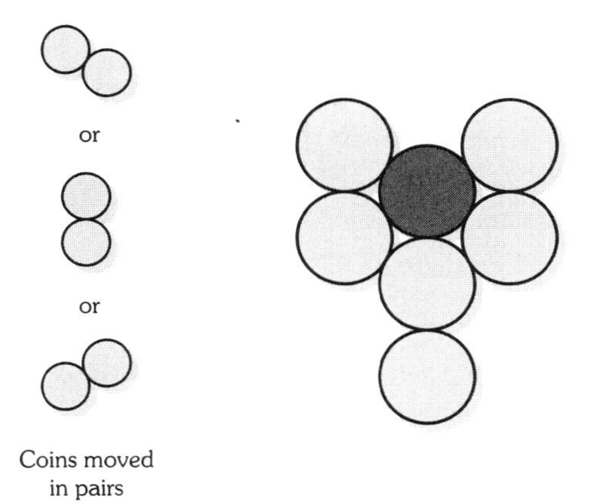

Moving one pair of adjacent coins at a time, form the Y in another place so that the central, darker coin is out of the Y. "Out" means that it does not touch the Y at all. Can you do this in five pair-coin moves? Note that you are not allowed to move the central, darker coin!

Solution on page 65.

Broken Square

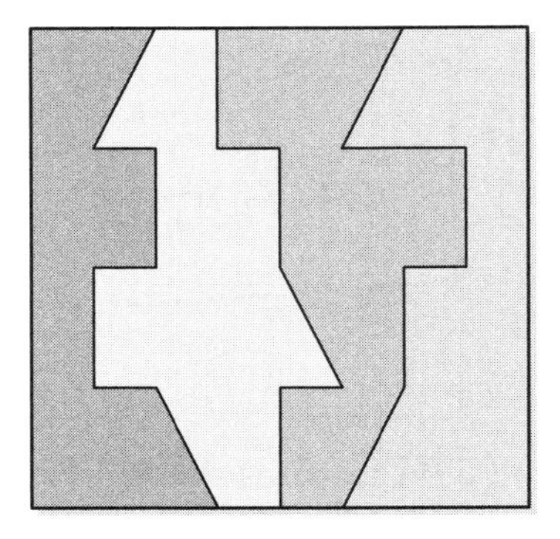

The square is broken into four parts. Two of them have the same area. Determine which ones these are.

Solution on page 65.

Why Tee Tee

The following puzzle was aired on a local radio station as a contest: "There is a well-known sequence represented by the three letters YTT. There are several ways in which these letters can be replaced with another three letters so that it is still true, but only one way so that the new sequence looks similar to that starting one; it may be something like BAA. When can this happen?" Knowing that the answer to the radio contest was, "It can happen two days after this aired," the question is, On which day was this puzzle aired on the radio?

Solution on page 65.

Elliptic Proportions

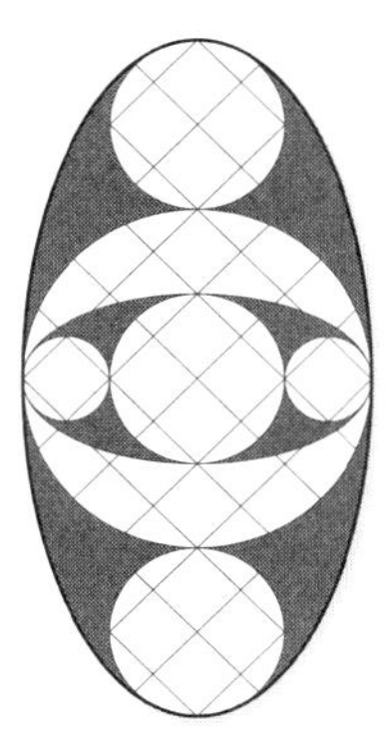

In the elliptic shape, what is the ratio of the shaded area to the unshaded area? The grid lines of the diagram do not count.

Solution on page 66.

Netting

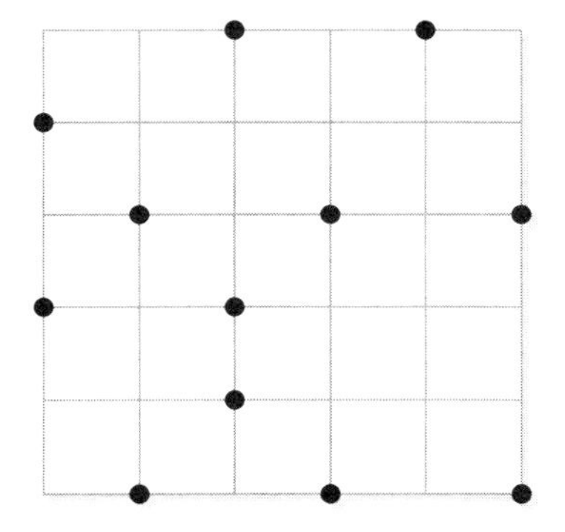

Link all twelve dots on the 5×5 grid with a chain of eleven consecutive connected straight segments. All segments must have exactly the same length and can touch dots only at their ends. Segments may cross one another.

Solution on page 66.

In the Right Triangle

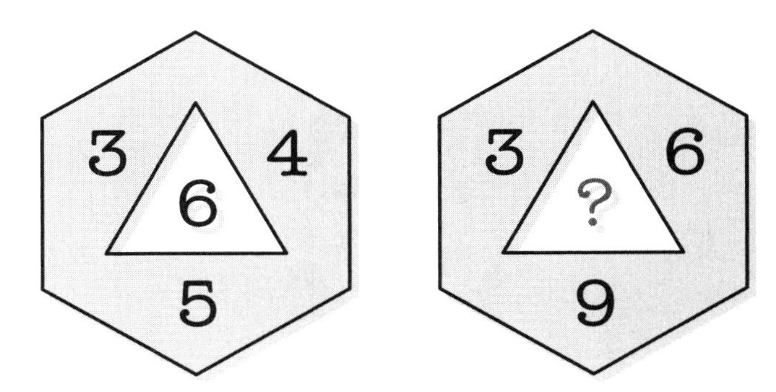

What one-digit number should replace the question mark in the triangle on the right?

Solution on page 67.

Checkered Challenge 58

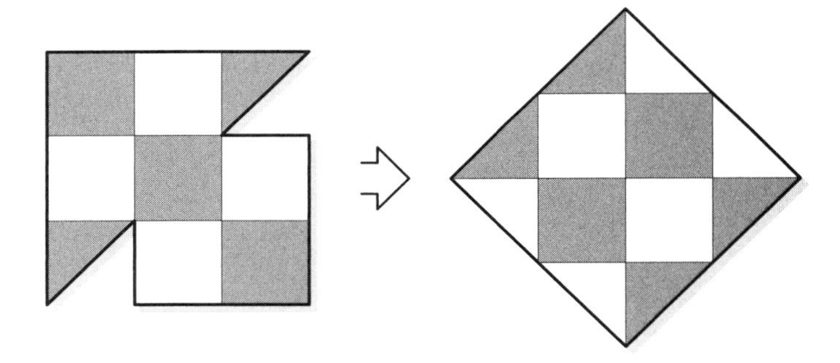

Divide the checkered arrow-like shape on the left into four pieces so that the pieces can be rearranged to form the checkered square on the right.

Solution on page 67.

Two Unicursal Bricks

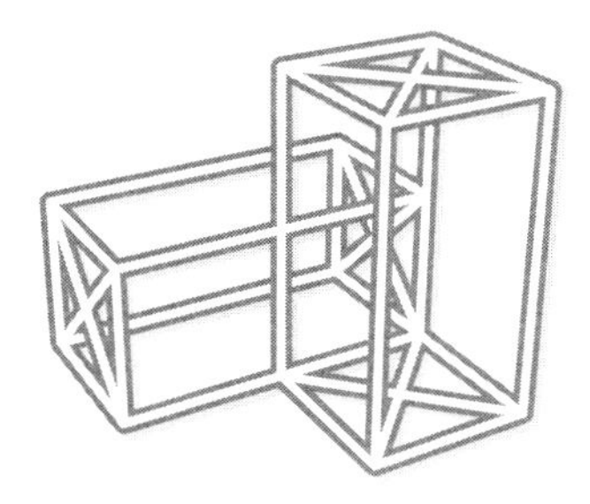

Draw one continuous, open-ended line within the three-dimensional space defined by the contour of the two-brick shape in the illustration. The line cannot cross itself, and no edge of the shape should be passed twice.

Solution on page 68.

Profiles: The Solids

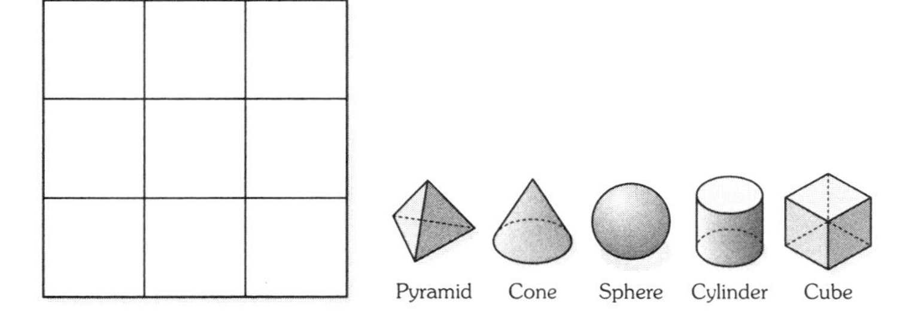

Each solid shares a profile, either vertical or horizontal, with at least one of the other solids. The pyramid shares a profile with the cone, a triangle. The cone, the sphere, and the cylinder all share a profile, a circle. Moreover, the cylinder shares a profile with the cube, a square. Now, place all of the five solids in the 3 × 3 grid (one solid per box) observing the following rule: "If a row, column, or a main diagonal contains more than one solid, all these solids must share the same profile within that particular line."

Solution on page 68.

The Wrench Dividing

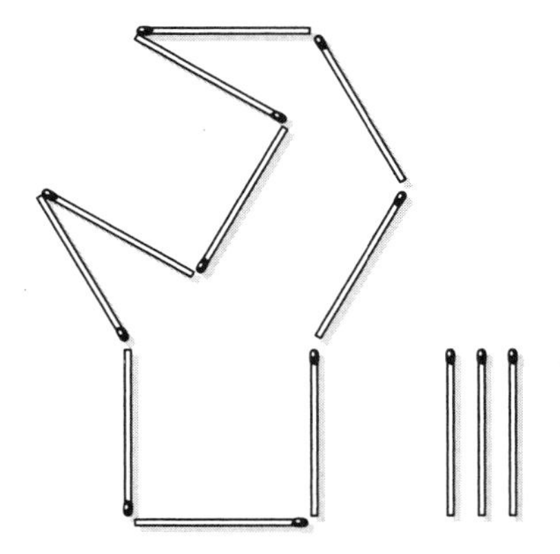

Using three additional matchsticks, divide the wrench into two parts of exactly the same area.

Solution on page 68.

In the Same Plane

Five pencils are put on the table as shown. Which two of them are in the same plane?

Solution on page 69.

The Triangle Quest

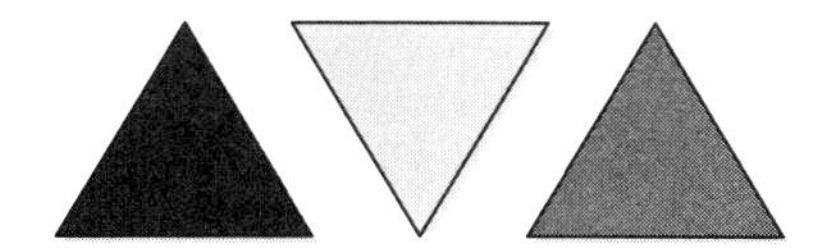

What is the maximal number of equilateral triangle outlines that can be formed using the three congruent equilateral triangles? An outline counts only if it forms an equilateral, unbroken triangle, no matter what size it is. You can rotate, move, and overlap triangles as you wish, but are not allowed to break, cut, or bend them. Note that all three triangles are opaque, not transparent.

Solution on page 69.

A Smooth Way through the Maze

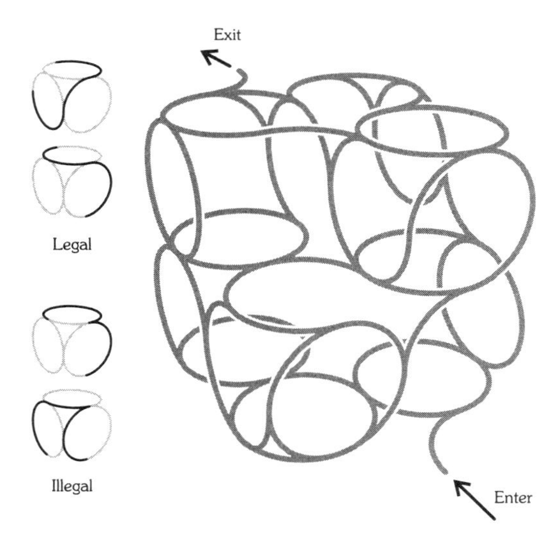

Find a smooth way (without sharp turns) through the contour maze. Examples of legal and illegal turns in the path are shown in the small illustrations next to the maze.

Solution on page 69.

Elastic Trios

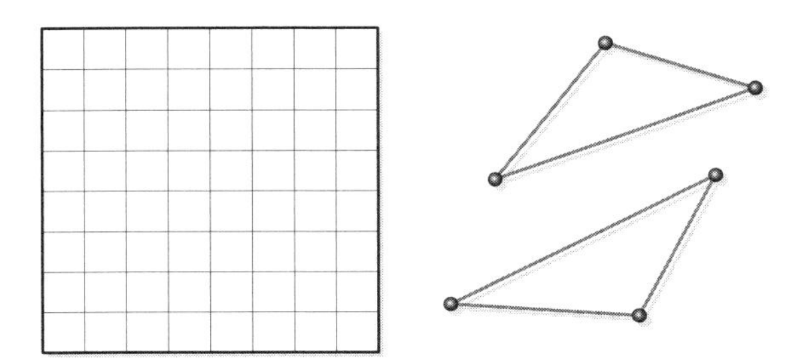

You have two elastic necklaces with three small beads each. Put all six beads on the perimeter of the grid so the segments of elastic thread divide the area of the grid into four parts of the same area. No two beads can share the same point, while the segments of elastic thread can cross each other.

Solution on page 70.

The Tube Colors

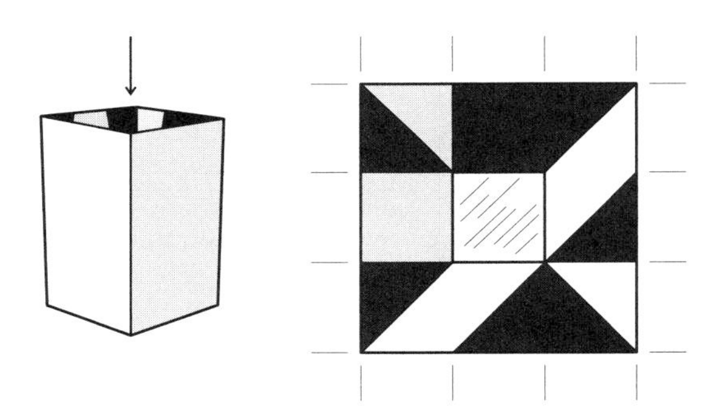

The square tube on the left has some black-and-white pattern inside it. Looking through the tube from the top along the arrow, one sees the pattern in perspective on the right. Can you determine whether the total area of all the black parts is larger, the same, or smaller than the total area of all the light parts? The lines of the diagram do not count.

Solution on page 70.

The Air Bubble Challenge

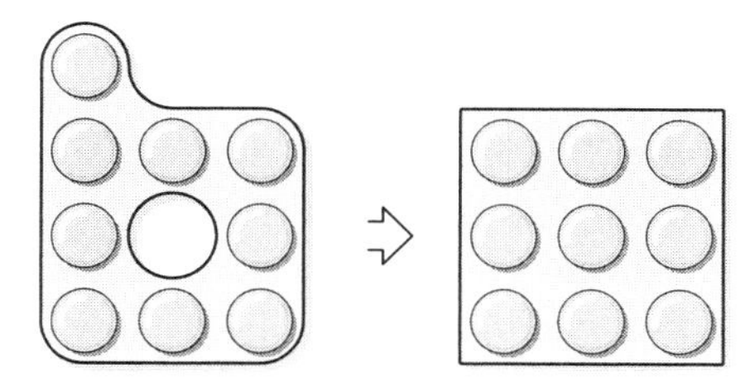

Divide the b-like shape (with rounded corners) that was cut out of a sheet of bubble wrap into four parts that can be rearranged into the square on the right. You are not allowed to make cuts over air bubbles, and every part must contain at least one of them. Also, no part can be flipped over.

Solution on page 70.

Three Fragments

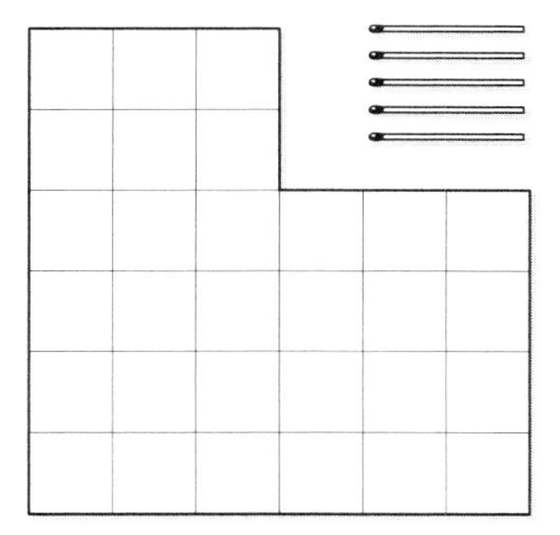

Using five matchsticks, divide the shape into three parts of the same area. You can put matchsticks only along the lines of the grid in the shape. Note that a matchstick is as long as two small boxes of the grid.

Solution on page 71.

An Odd Field

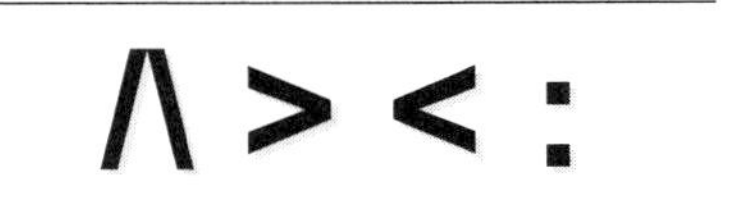

An applicant was filling out an application form. He had no significant problems with all the fields of the application form except one, which happened to be coded in an odd way—just as shown in the illustration. After a moment of hesitation, the applicant figured it out and filled in this field correctly too. What did the applicant write in the field? (Hint: The field was not NEW to the applicant.)

Solution on page 71.

Finding the Solution

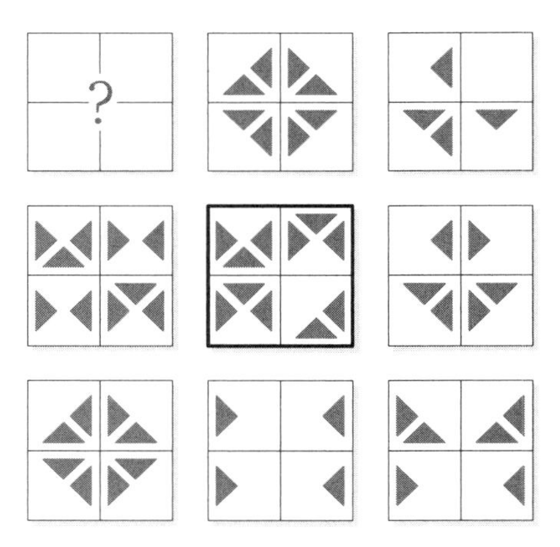

Can you find the pattern that should be drawn in the top left box instead of the question mark to complete the sequence? The key to the solution is shown in the central pattern.

Solution on page 71.

Pentatriangles

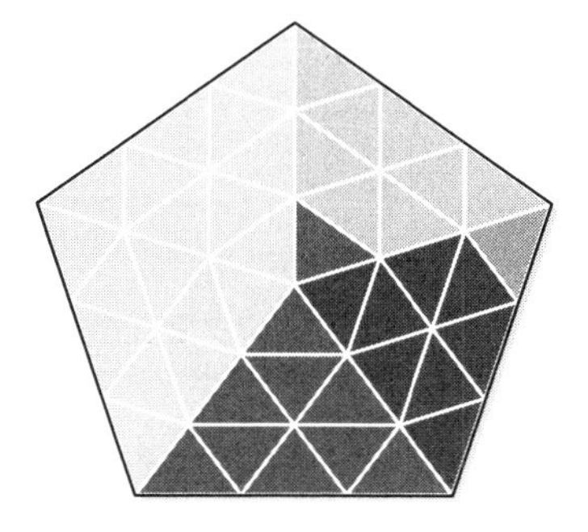

The three-dimensional pentagonal cap has five equilateral triangle faces. It also has two areas, lighter and darker. First, divide the lighter area into three parts of equal area and shape (when unfolded into flat shapes). Then do the same with the darker area of the cap.

Solution on page 72.

The Drop

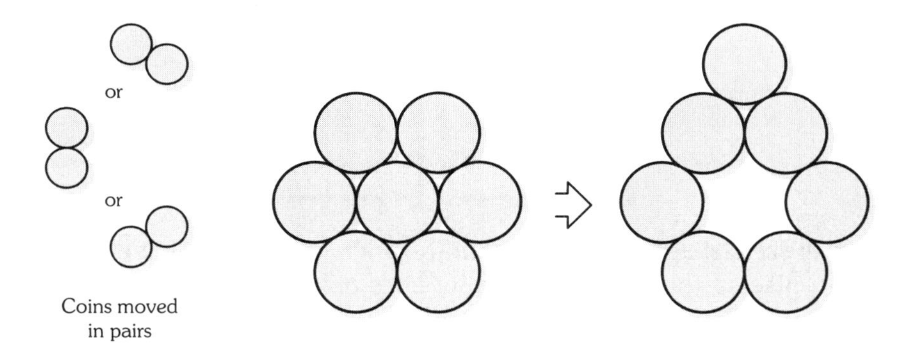

Transform the seven-coin hexagon on the left into the "drop" frame on the right by moving one pair of adjacent coins at a time (see small sample diagrams). Can you achieve this in four pair-moves?

Solution on page 72.

Triangle & Two Matchsticks

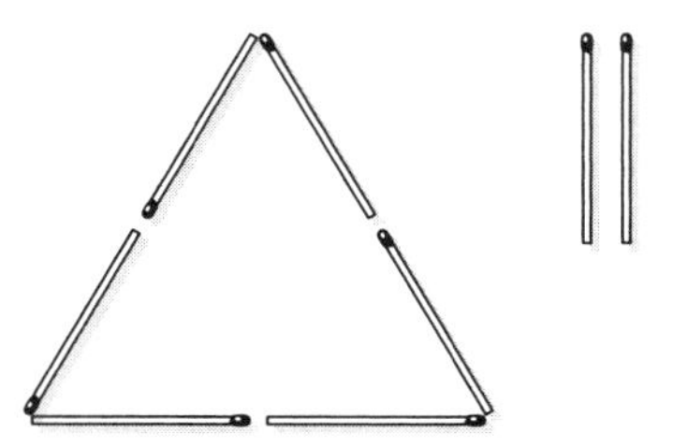

Using two additional matchsticks, divide the equilateral triangle into two parts of exactly the same area.

Solution on page 72.

Easy L-Packing

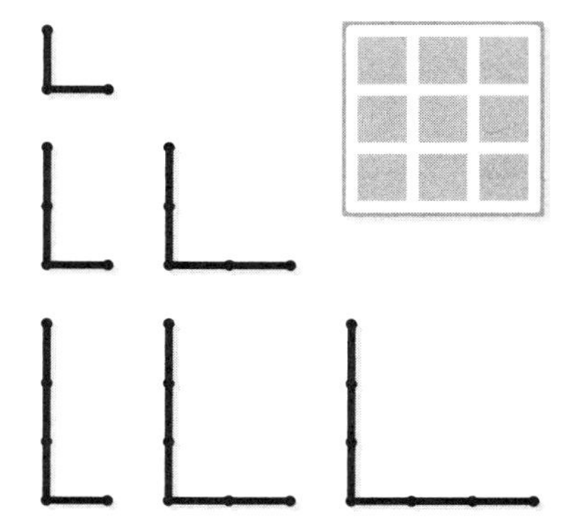

Put all six L-shapes entirely within the 3 × 3 square board at the top right, placing them along the white lines only. L-shapes can be rotated and flipped over, they can touch and cross each other at the dots, but no straight segments can overlap. Also, note that L-shapes are rigid and cannot be bent or folded.

Solution on page 73.

Golden Budget

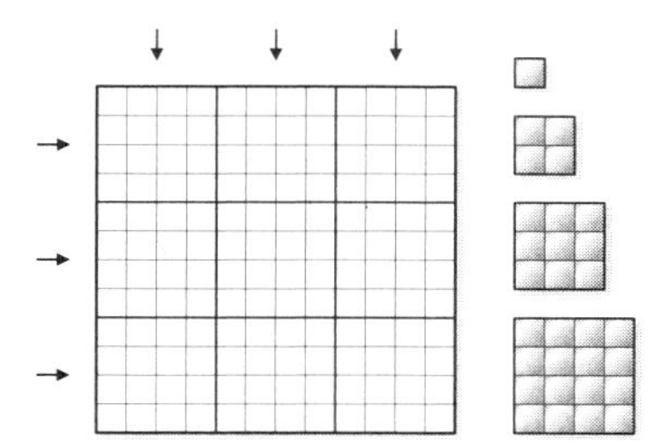

Put the four golden square bricks with area proportions of 1:4:9:16 onto the 3 × 3 grid on the left so that each of the nine boxes contains some portion of gold in it, and each row and column contains the same amount of gold. You can put the bricks onto the grid as you wish, but you are not allowed to overlap them, damage them, turn them on their side, or put a portion of a brick outside the grid's outline. A brick can occupy several boxes, but what portion of it belongs to which box is defined by the outline of each particular box.

Solution on page 73.

The Shape-Color Connection

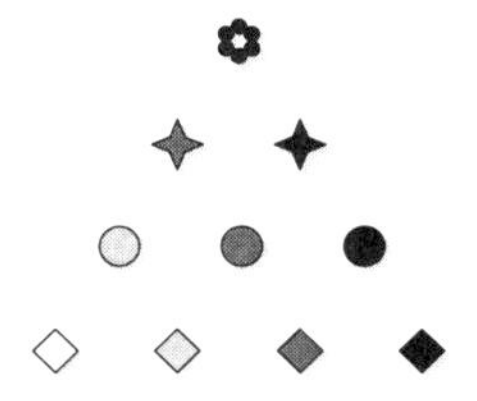

Link the ten colored shapes in the triangle with exactly seven connected straight lines. On your route any two consecutive figures must differ both in shape and in color, and each figure must be visited only once. Lines must go through the centers of the shapes and may cross each other, but never at a shape. "Going through the center" includes passing straight through a shape and turning at a shape's center.

Solution on page 73.

Separate the Shapes

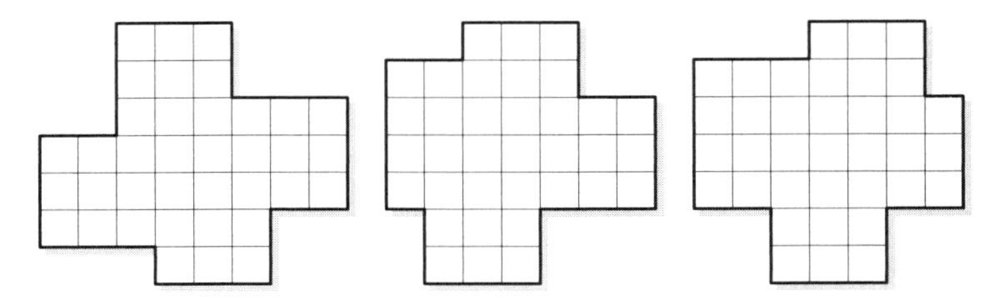

Each of the shapes is assembled from the same set of two pieces. While forming these shapes, the pieces were rotated, turned over, and even overlapped. Figure out the exact shape of the two pieces and how they were arranged to form each shape.

Solution on page 74.

S-Days and T-Days

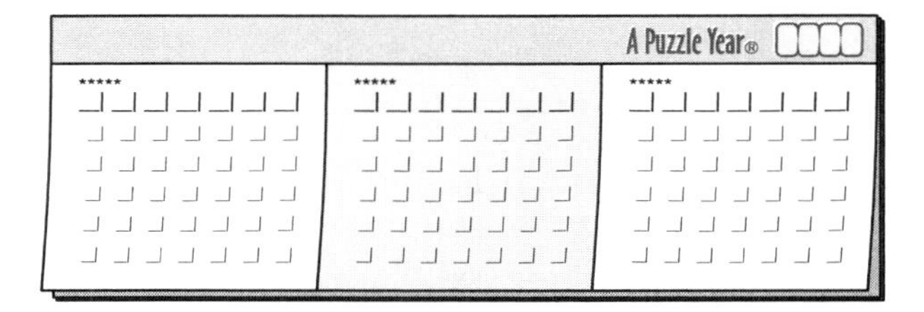

Well after the winter holidays, on the first day of a month, two friends made some curious discoveries about the calendar on the wall:

"Look, the number of the days in this month that start with 'T' is equal to the number of the days that start with 'S,'" one of them said.

"Cool!" replied the other, looking at the calendar. "But the same is true about the previous month as well!"

"Hmm...And the total number of these days is the same for each of these months," added his friend.

Now, can you determine precisely when (the exact day of the week and the month) the conversation between the two friends took place?

Solution on page 74.

The Puzzling Cross

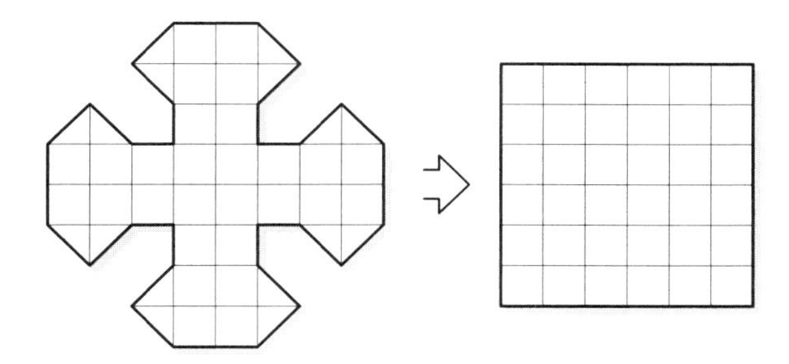

Divide the cross on the left into six parts so that the parts can be rearranged into the square on the right.

Solution on page 74.

Round-Up

The square is made of four parts. Two of them have the same area. Which ones are these?

Solution on page 75.

The Square-Triangle Couple

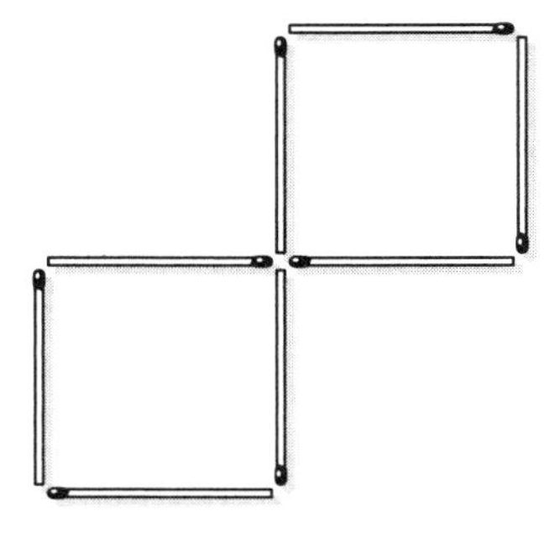

Eight matchsticks are arranged into two congruent squares. Can you move exactly four matchsticks to get exactly two congruent triangles instead of the squares?

Solution on page 75.

The Die Stack

You have two standard dice (the values that sum to seven on opposite faces) with truncated corners. Now there are two puzzles. Put one die on the other in a small stack so that the sum of the pips you can see from the point directly above the stack (Top Sum) is greater than the number of pips on the face that touches the table (Bottom Sum).

Puzzle 1. Top Sum is 8 times greater than Bottom Sum.
Puzzle 2. Top Sum is 12 times greater than Bottom Sum.

Solution on page 75.

T-Unicursal

Using all seven pieces of the classic Tangram, form the T-shape on the right. The whole shape must be assembled so that its pattern created with all the pieces' outlines is unicursal. This means that you should be able to draw it in one continuous, open-ended line that does not cross itself. The pieces can be rotated and flipped, but not overlapped. Note that when pieces touch each other along their edges they form a single line.

Solution on page 76.

The Square Quest

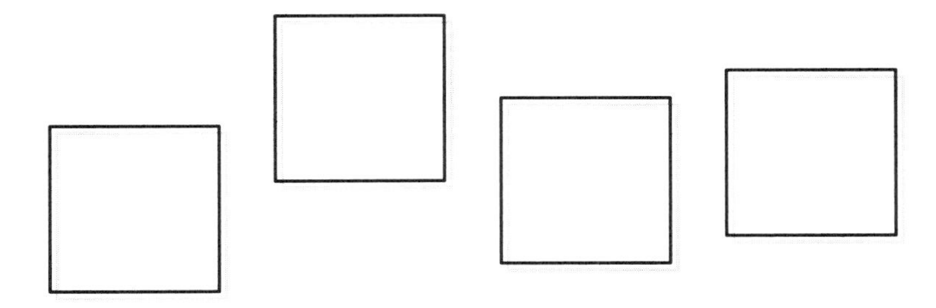

What is the maximal number of square outlines that can be formed on the plane using the four congruent (transparent) square frames? An outline counts only when it is a perfect, unbroken square of any size. Outlines can cross and overlap each other. You can rotate, move, and overlap the frames as you wish, but you are not allowed to break, cut, or bend them.

Solution on page 76.

Signs & Symbols

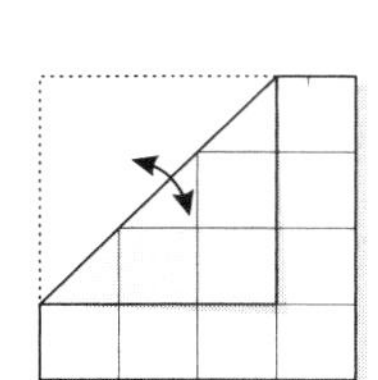

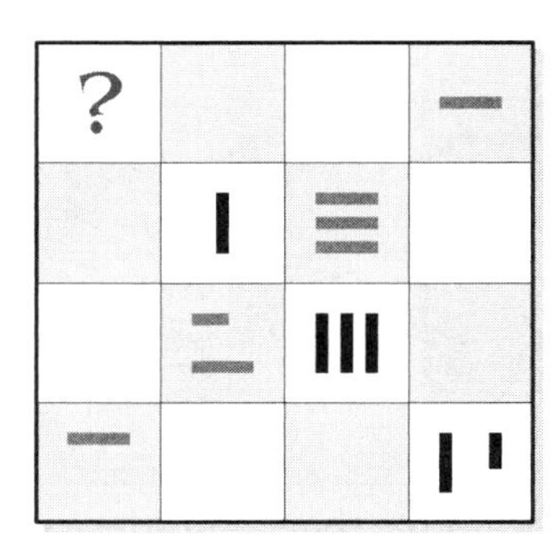

Which symbol should be placed in the top left cell of the checkered 4 × 4 paper square at the right? (Hint: The small folding diagram (on the left) can help you understand the idea of how the symbols were created and distributed within the checkered square.)

Solution on page 76.

1-2-3 Transforming Puzzle

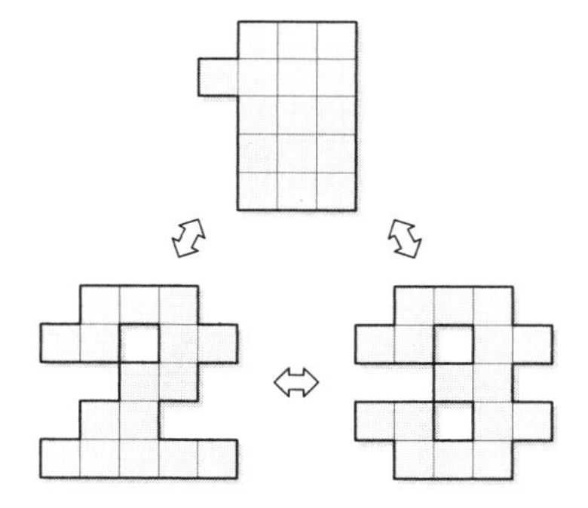

Cutting only along the lines of the grids of the digits, divide each into four pieces so that the set of pieces will be the same for each digit. In other words, using this set you can form the 1, then the 2, and, finally, the 3. Remember that you can rotate pieces as you want, but not flip or overlap them.

Solution on page 77.

The Viewpoint

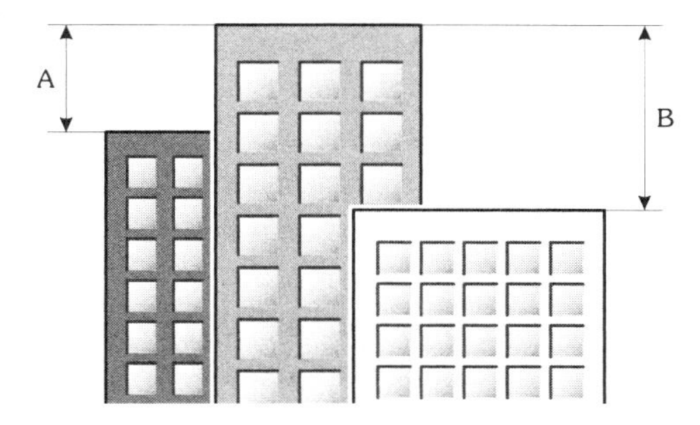

Standing right beside the window in his office, SGJ sees the facades of three buildings as shown in the illustration. The question is, What would happen to imaginary segments A and B if SGJ lowers his viewpoint by sitting down in an armchair? Would each get longer, get shorter, or remain the same?

Solution on page 77.

Play with an Erasure

Write the word "shake" as shown. Then erase a part of the "h," and you have another word: "snake." How many such word pairs can you find? Samples of the letters of the English alphabet are provided for your convenience.

Solution on page 77.

Trapezoid Contours

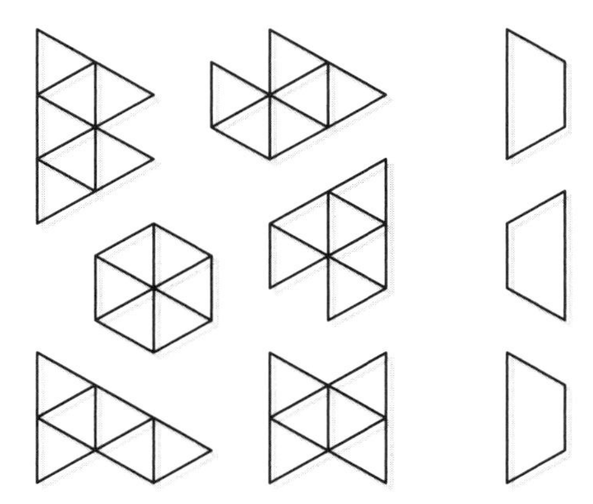

Determine which of the six contour patterns can be formed using the three single (transparent) trapezoid contours on the right. The trapezoids can be rotated and overlapped.

Solution on page 78.

Add a Row

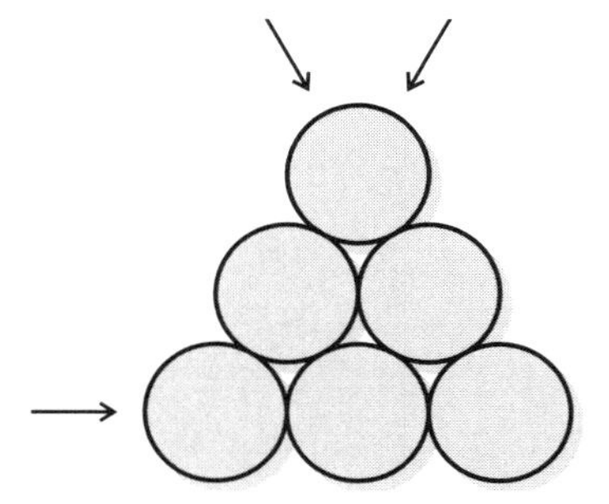

In the coin triangle you can count three straight rows containing three coins each. Now, move two coins into new positions so that you can count four straight rows of three coins each.

Solution on page 78.

The M-Shaped Count

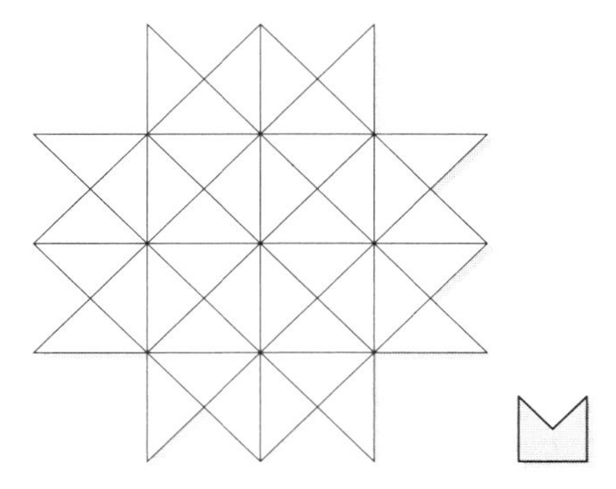

How many mitre shapes (M-shapes) of all possible sizes and orientations are in the grid? All M-shapes must be similar to the small illustration next to the grid. In fact, a mitre is a square sans a triangular quarter.

Solution on page 78.

No Magic

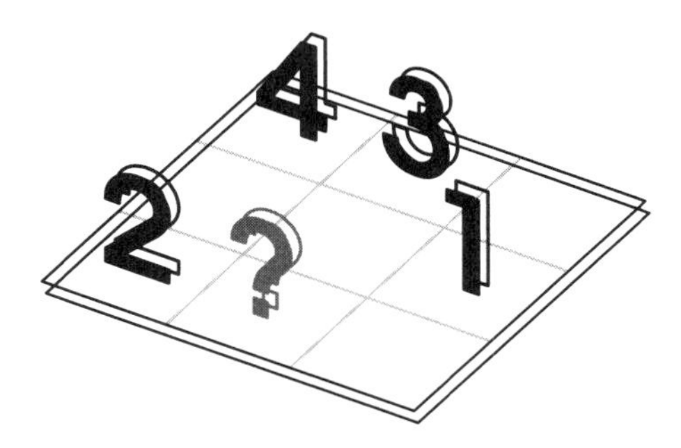

Which digit should go instead of the question mark in the 3 × 3 board? Note that the answer is not 5, and it differs from those already shown.

Solution on page 79.

What Dice It Matter?

The five dice in the cross are identical, including the orientations of the pips on their faces. The dice are standard, so the values on opposite faces always add up to 7. Dice touch each other with the same numbers so that all pips on the touching faces exactly match. In other words, each pip touches exactly another pip. Determine how many pips should be drawn on the two faces with the question marks on them, and show their exact orientation.

Solution on page 79.

Hexa Differing

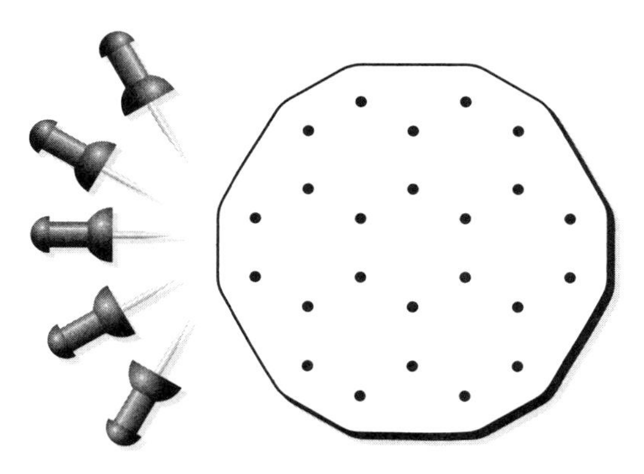

Place five push-pins in five holes in the board so there are no two pairs of equidistant push-pins.

Solution on page 79.

The Heart of the Match

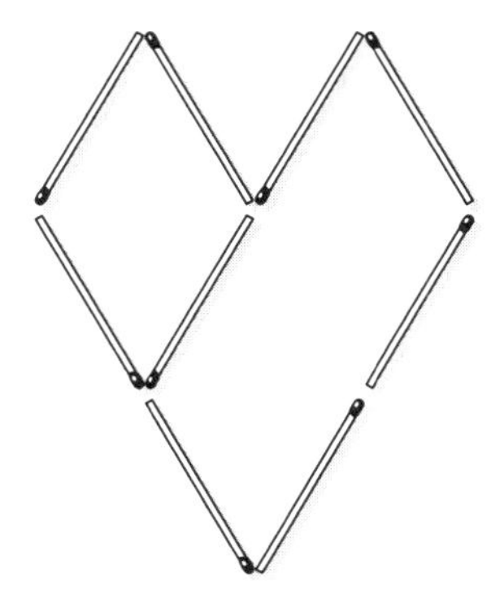

This shape is divided into two different parts. It is the start shape for the three following puzzles.

Puzzle 1.	Move two matchsticks to form a shape divided into exactly two congruent parts.
Puzzle 2.	Move three matchsticks to form a shape divided into exactly three congruent parts.
Puzzle 3.	Move four matchsticks to form a shape divided into exactly four congruent parts.

Solution on page 80.

DigitCount

What digit should replace the question mark?

Solution on page 80.

The Book Staircase

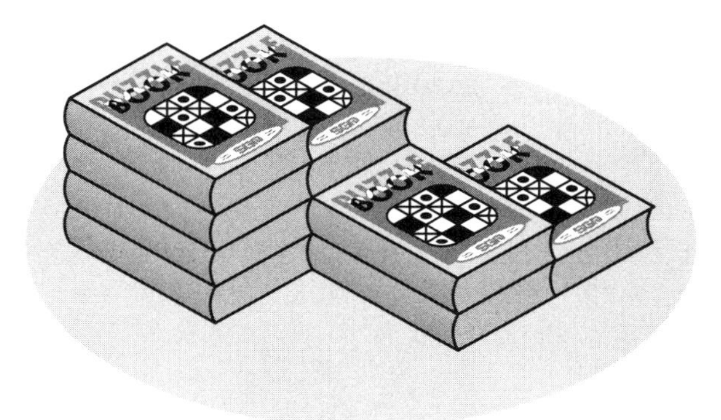

Add two straight lines to the books so that you can see some additional books similar to those shown.

Solution on page 80.

Two Different Triangles

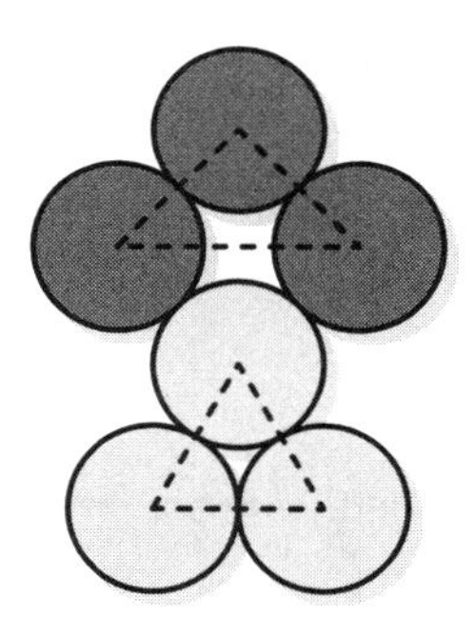

The coin shape consists of two triangles. One of them, formed of "heads-up" coins, is equilateral, while another, formed of "tails-up" coins, is isosceles. Observing the "double-touch" rule and using the least number of single-coin moves, rearrange the "heads-up" coins into an isosceles triangle, and the "tails-up" coins into an equilateral triangle.

Solution on page 81.

Inside the Grid

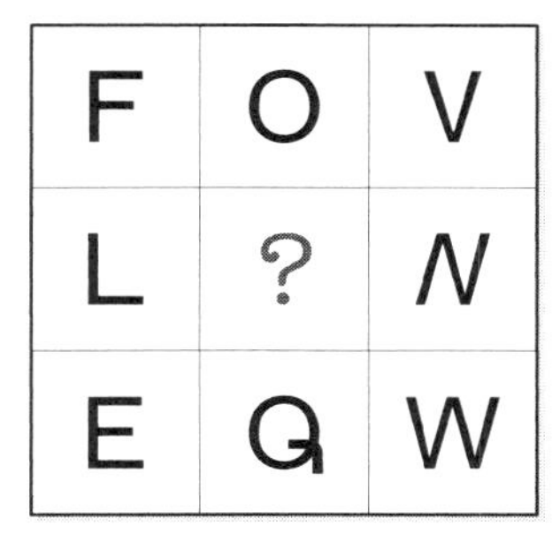

What letter should replace the question mark in the grid?

Solution on page 81.

Drop & Match

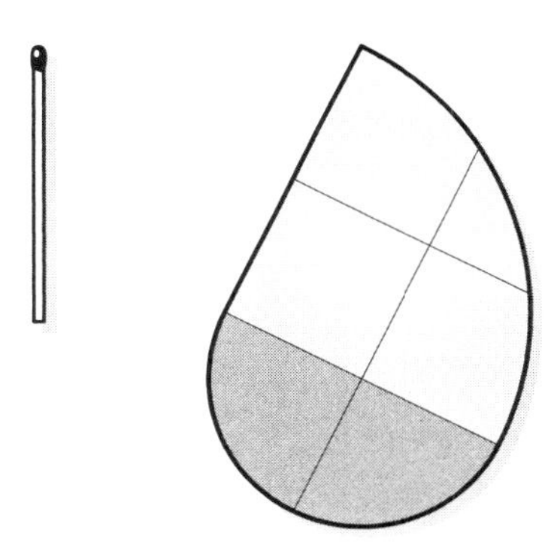

With a single matchstick, divide the drop into two parts of the same area. Note that the two shaded parts of the drop are sectors of two circles with radii 1 and 2 units, respectively. The matchstick is 2 units long.

Solution on page 82.

Mag^{2}netic

You have a flat square that is a strong magnet. Also, you have a big stock of equal coins with diameter equal to the square's side.

What is the maximal number of coins that you can affix directly to the surface of the square, on both of the square's sides? Coins will be held by the magnet when they touch the square's faces (not just its edges or corners) directly with some real (even very small) area of their sides, but never with their rims. Coins can touch each other, but not overlap.

Solution on page 82.

The Antique Ring

This metal ring looks to be slightly damaged, since in one of its openings (marked with a question mark) a wire element is missing. Can you restore that element?

Solution on page 82.

The Factory Block Puzzle

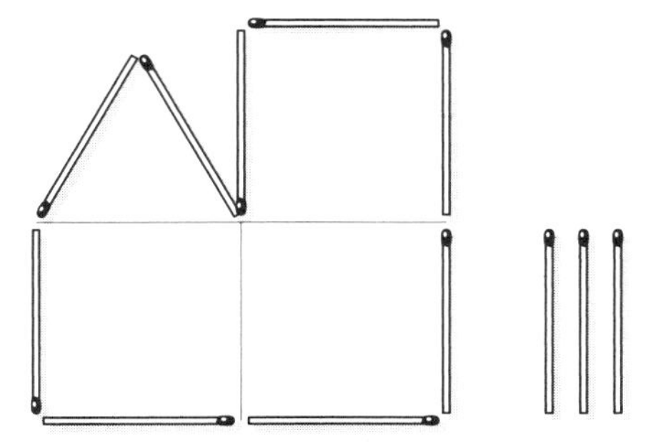

With three matchsticks, divide the factory block into two parts of the same area.

Solution on page 83.

Puzzle Card

The 16-digit number of a Puzzle Card credit card is ciphered in 16 vertical stripes as shown. Can you decipher it?

Solution on page 83.

Coin Upside-Down

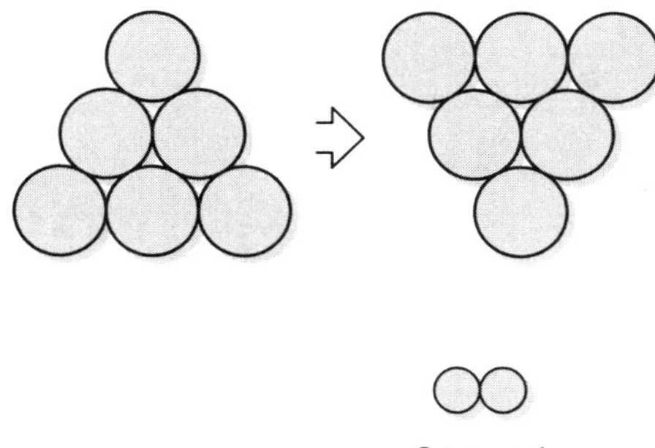

Coins moved
in horizontal pairs

Six identical coins are arranged into a triangle as on the left. Making only "horizontal-pair" moves, and observing the "double-touch" rule, turn the triangle upside down as on the right. Achieve the goal in the least number of the moves. A "horizontal-pair" move is an orthogonal slide of any two horizontally adjacent coins as shown in the small diagram beneath the goal position.

Solution on page 84.

The Puzzle Infinity

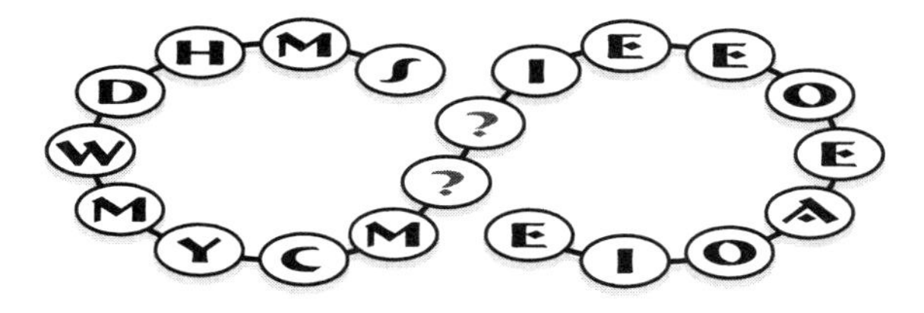

Which two letters should replace the question marks in the central circles of the diagram?

Solution on page 84.

The Caravel

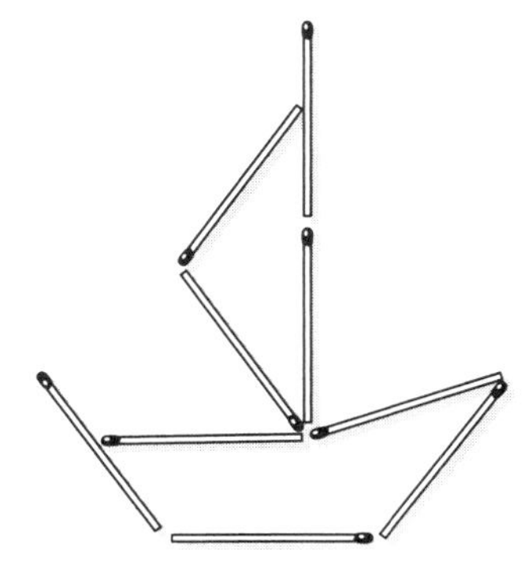

Move five matchsticks so that the caravel of exactly the same shape sails in another direction. Note that in this puzzle loose ends of matchsticks are allowed.

Solution on page 85.

Twin Cubism

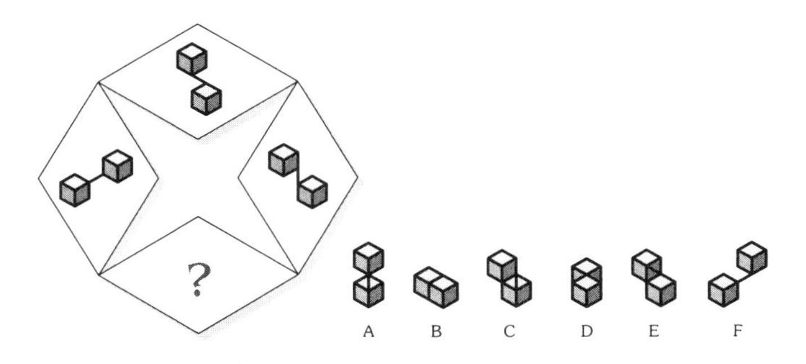

Which of the six shapes (A–F) should replace the question mark in the diagram?

Solution on page 85.

Three L's to a T

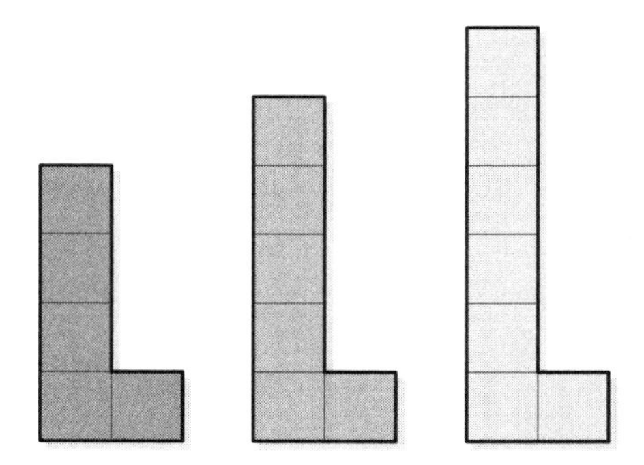

Using the three L's, assemble a symmetric capital T. It should have all of its legs straight, something like this: T. You can rotate and flip pieces as you wish, but no overlapping is allowed.

Solution on page 85.

Equal Perimeters

The square is divided into five parts. Which two have the same perimeter?

Solution on page 86.

What's There in the Square?

1		3
2	4	
	6	?

What number should replace the question mark in the grid?

Solution on page 86.

G-Knights Exchange

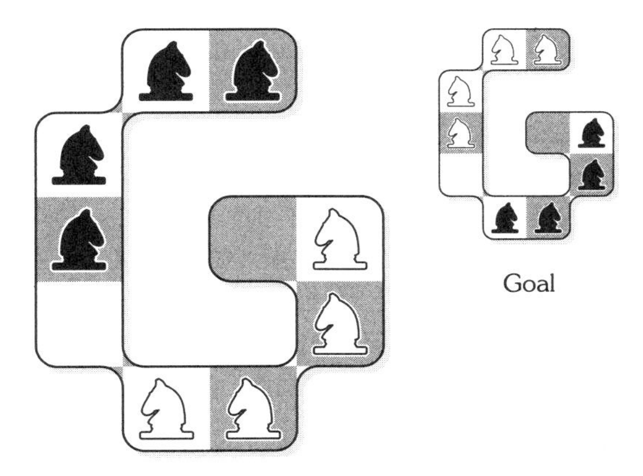

Eight chess knights, four white and four black, are placed on the special G-chessboard as shown in the big diagram. Now, using normal knight moves, exchange the white and black knights as shown in the Goal diagram. Counting a consecutive series of leaps by one knight as one move, can you exchange the knights in exactly 11 moves?

Solution on page 87.

Rectangle Differing

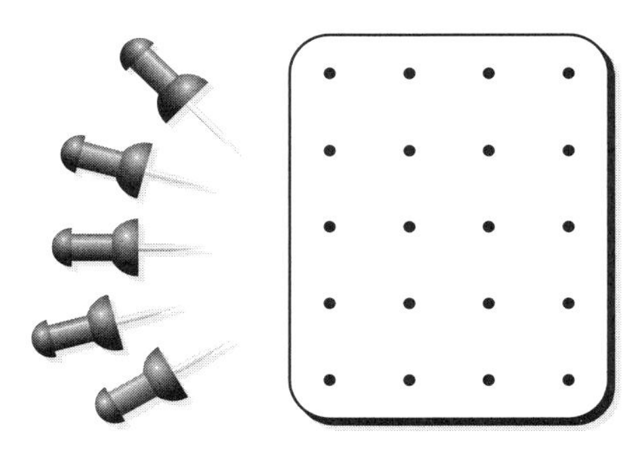

Place five push-pins in five holes in the board so that there are no two pairs of equidistant push-pins.

Solution on page 87.

Four More Triangles

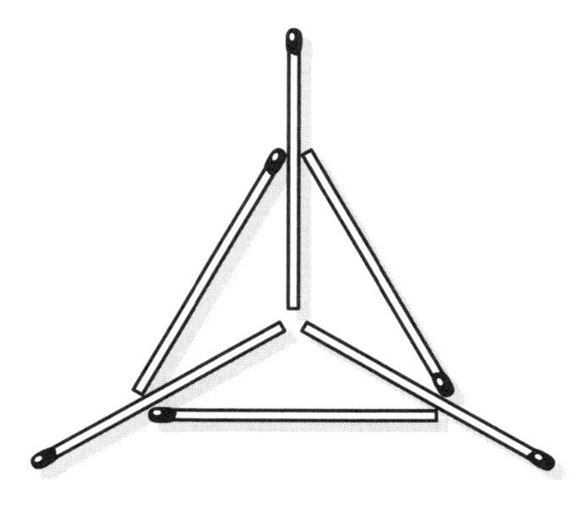

There are four triangles in this figure, counting ones of all sizes. Move a minimal number of matchsticks to make eight triangles of any sizes. Note that in this puzzle loose ends of matchsticks are allowed.

Solution on page 87.

The Four Snakes Puzzle

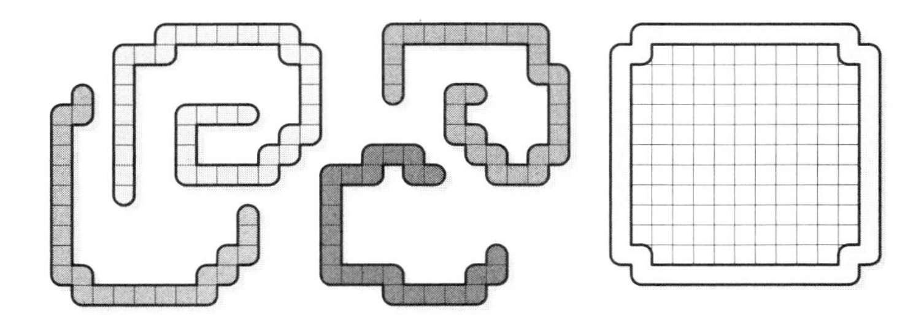

A square yard in the Quince Palace has four paths paved as four different snakes. The shapes of these paths are given, so the only challenge is to put them within the yard. Paths can be rotated and overturned, but do not overlap them.

Solution on page 88.

Black or White in Cube

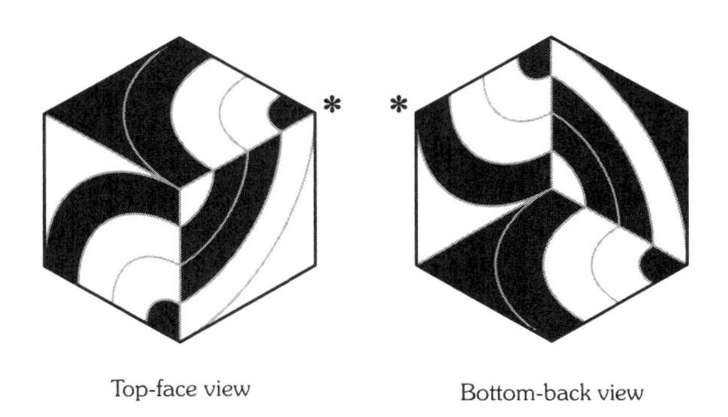

Top-face view   Bottom-back view

The surface of a cube is divided into black and white parts as shown. Determine the ratio of all the black parts to all the white parts. The gray dividing lines on the cube's faces do not count.

Solution on page 88.

Pentomino Switch

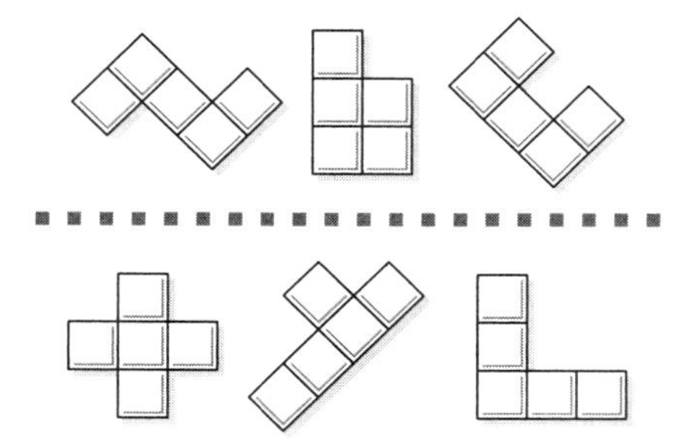

When two pentominoes exchange places, a meaningful pattern is restored. Which two? You can rotate the pentominoes, but do not flip them.

Solution on page 89.

The Ancient Pyramid Puzzle

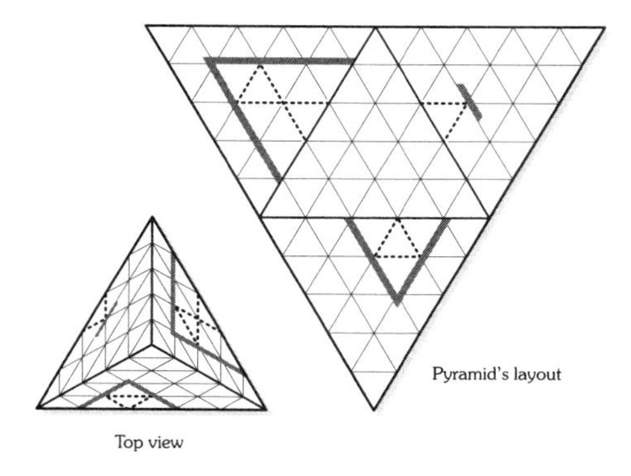

Each face of the pyramid on the left has a pattern formed of thick and dashed lines. The three faces and their patterns are shown in the top view of the pyramid, and in its layout unfolded next to the pyramid. What pattern should be on the bottom face of the pyramid? Bear in mind that all patterns' lines must run along the grid's lines covering the pyramid.

Solution on page 89.

Broken Watch

The hour hand and the minute hand of the wristwatch are not synchronized properly. To fix the problem, divide the clock-face into three parts that can then be reassembled so that both hands are perfectly synchronized. Your cuts must be along the white lines on the clock-face.

Solution on page 90.

Coins Apart

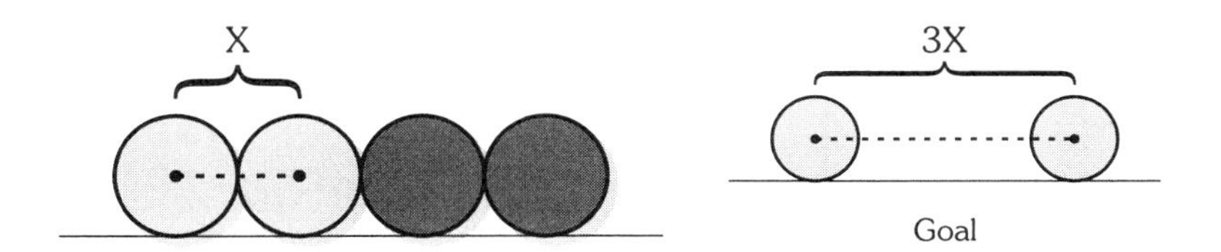

Four equal coins, two "heads-up" and two "tails-up" ones, are placed in a horizontal line. The distance between the centers of the two "heads-up" coins equals the width of a coin, X.

The object is to reach, in as few sliding-coin moves as possible, a position for the "heads-up" coins so that the distance between their centers equals 3X, as shown in the Goal diagram. Observe the "double-touch" rule. Additionally, at the end both "heads-up" coins should be along the same horizontal line shown.

Solution on page 90.

The Deer Puzzle

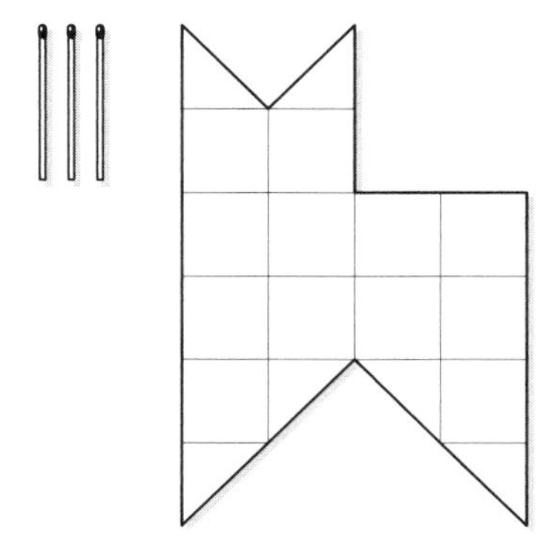

Using three matchsticks, divide the shape into three parts of the same area. Note that a matchstick is as long as two small boxes of the shape's grid.

Solution on page 91.

Cheap Victory

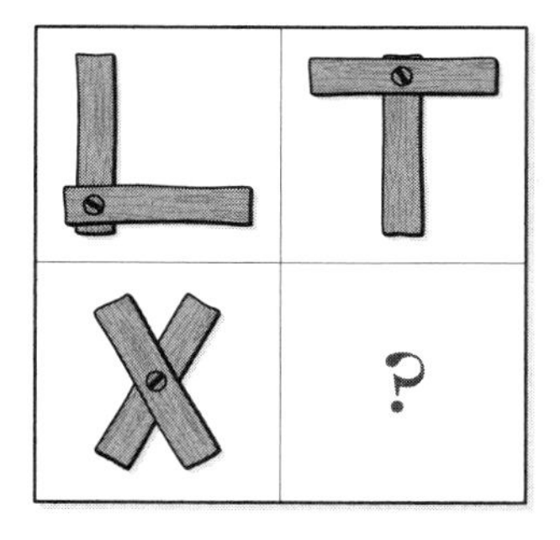

Which letter should replace the question mark?

Solution on page 91.

Down the Street or… Finding the House

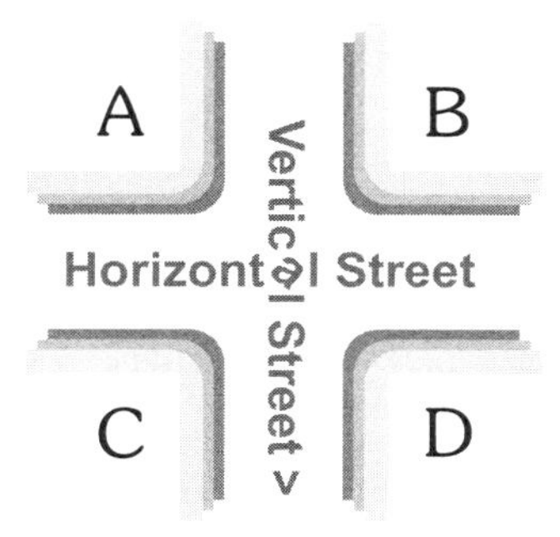

Vertical Street, which starts at the north and runs to the south, crosses Horizontal Street (its direction is not revealed) at a right angle, as shown in the illustration. There are exactly nine houses on each street, and the sum of house numbers in blocks A and D equals the sum of house numbers in blocks B and C. As you move along the street from beginning to end, house numbers are consecutive (1 through 9), and the odd-numbered houses are always on your left and the even-numbered houses on your right.

On which block is house #5 of Horizontal Street, if there are four houses in block B?

Solution on page 92.

97 Question 5

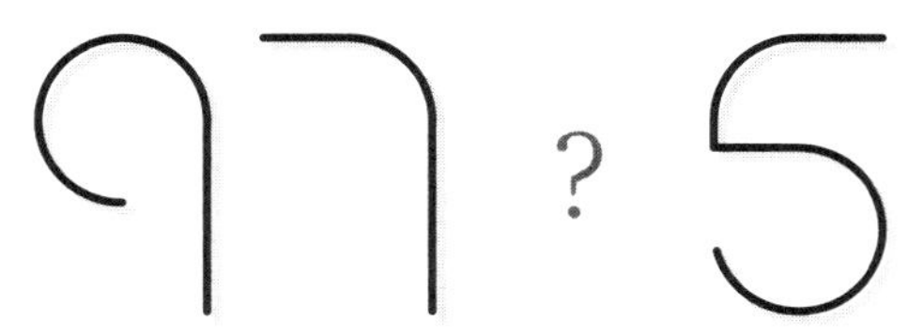

What digit should replace the question mark?

Solution on page 92.

Triangles & Digits

Place the four triangles on the right within the triangle frame so that all nine digits 1 through 9 appear; their shapes must be as shown under the triangles, but they can be rotated. Fragments of digits are depicted both on the triangles and on the frame. You can rotate triangles as you wish, but can neither overturn nor overlap them.

Solution on page 92.

Coin Invert

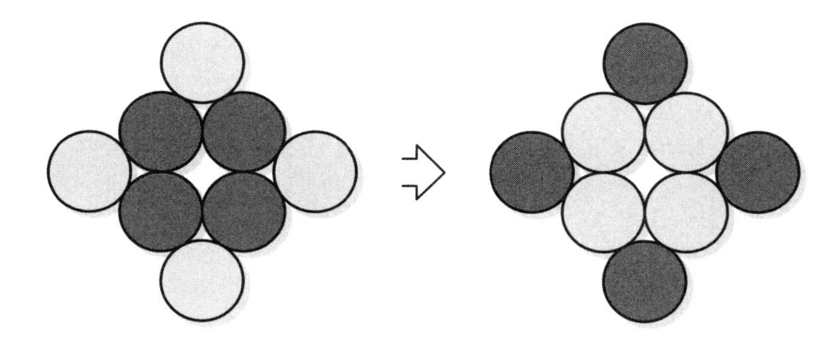

Eight equal coins, four "heads-up" and four "tails-up," form a star-like shape shown in the diagram on the left. Observing the "double-touch" rule, reach the position shown on the right within the minimal number of single-coin moves. The final orientation of the shape may differ from that shown.

Solution on page 93.

The Legendary Town

The picture shows a bird's-eye view of a legendary town, which is famous and desired as much as elusive and impossible to build. What is the name of that town?

Solution on page 93.

Cubism

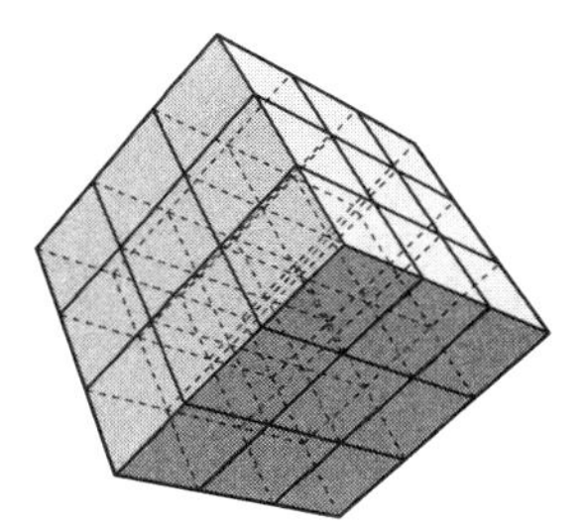

You have a $3 \times 3 \times 3$ cube of twenty-seven unit cubes.

Puzzle 1. Divide the cube into three coherent solid polyminoes with the same number of unit cubes such that their surfaces visible on the $3 \times 3 \times 3$ cube's faces are equal.

Puzzle 2. Divide the cube into three coherent solid polyminoes with the same number of unit cubes such that their total surface areas are equal. The polyminoes must differ in shape.

Solution on page 94.

The Matchstick Needle

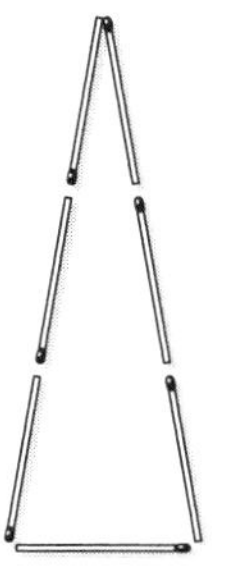

Move three matchsticks so that three triangles of the same shape and size appear.

Solution on page 94.

Shore Connecting

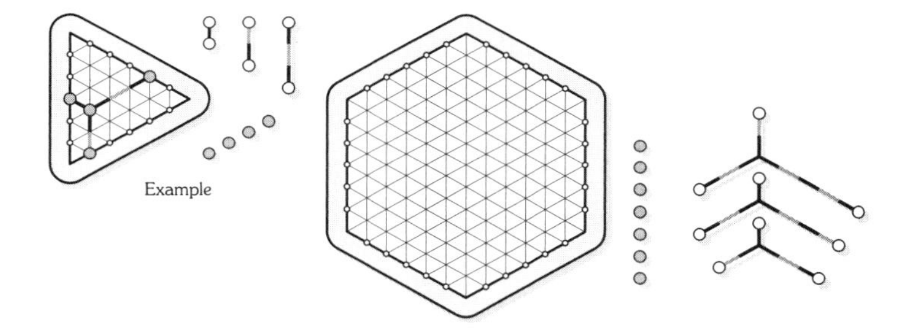

Build a network connecting all six sides of the hexagonal lake using the three bridges and seven single piers shown next to the lake. Place one pier per side of the lake and a single pier somewhere in the middle of the lake so that each end of each bridge is placed on one of these seven piers. You can rotate and flip bridges as you wish, but no free end of a bridge is allowed. Bridges must not overlap one another, except when their ends meet on the piers. A small example of the puzzle with a triangular lake, three simple straight bridges, and four single piers is shown in the small illustration on the left.

Solution on page 94.

Cube Differing

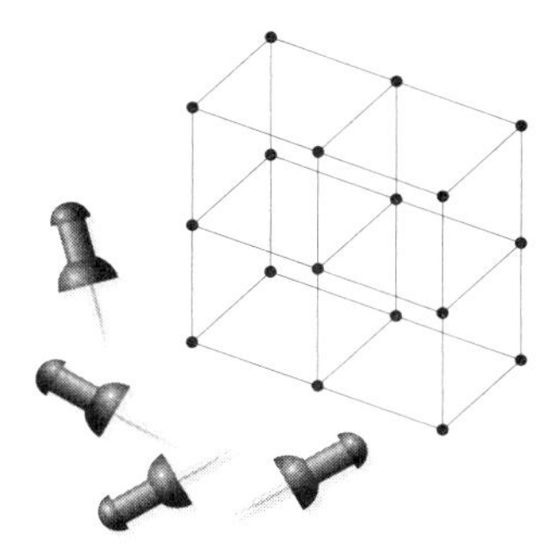

Place four push-pins in four nodes of the cubical lattice so that there are no two pairs of equidistant nodes with push-pins.

Solution on page 95.

The Right-Angle Framework

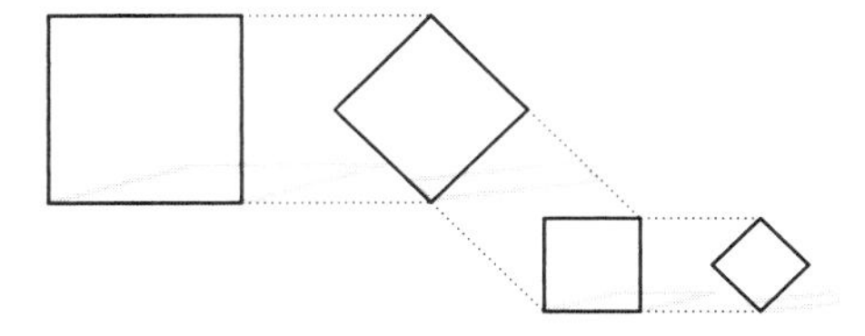

You have four contour (transparent) square frames. Note that each frame's diagonal is equal to the side of the previous frame.

Puzzle 1. What is the maximum number of perfect square outlines of any size that can be produced on the plane with these four square frames?

Puzzle 2. What is the maximum number of right isosceles triangle outlines of any size that can be produced on the plane with these four square frames?

For both puzzles you can rotate and overlap the frames as you need, but do not break, cut, or bend them.

Solution on page 95.

The VHS Tricky Packing

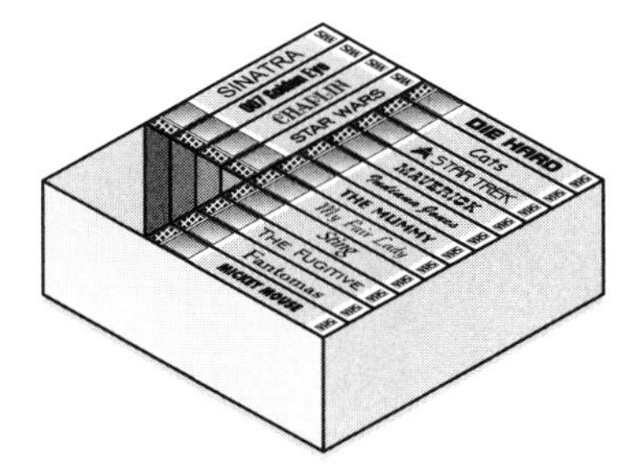

Fifteen VHS cassettes are placed in a $4 \times 11 \times 11$ box as shown. It is possible to add one more cassette to them and pack them all within the box again. How can it be done? Note that each cassette is $1 \times 4 \times 7$.

Solution on page 96.

Coin Triangle Theorem

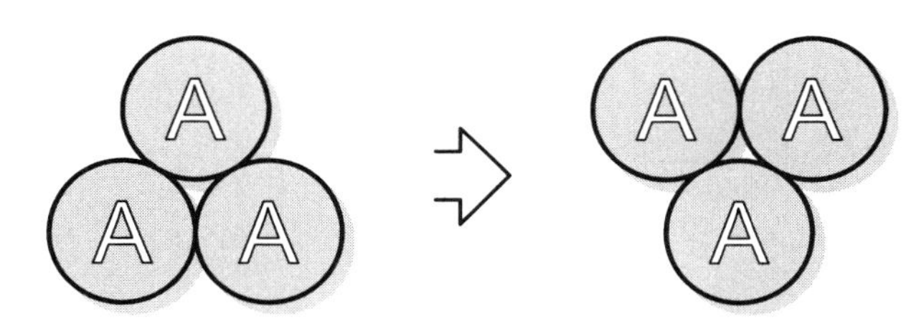

You have a triangle formed of three equal "heads-up" coins; their heads are marked with A's. The goal is to rearrange coins into another triangle as shown in the right diagram with some number of flipping moves. To perform a flipping move, choose a coin, flip it over, then move it to another position where it touches two unmoved coins. Can you prove whether it is possible or not to achieve the goal?

Solution on page 96.

Guess the Phone Number

phone: 406.729.?381

The phone number starts with 4 and ends with 1. It has an unusual property: when all the remaining digits but one are revealed it is possible to determine the missing digit, whatever its position is. In this case the seventh digit of the phone number is not revealed. What is this missing digit, if you know that it is not 5?

Solution on page 96.

Magnetic Tetrahedron

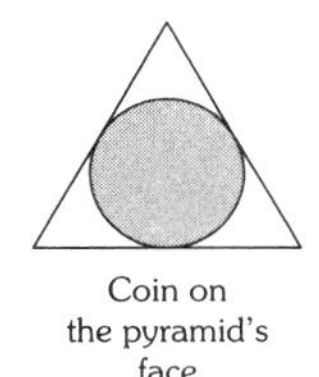

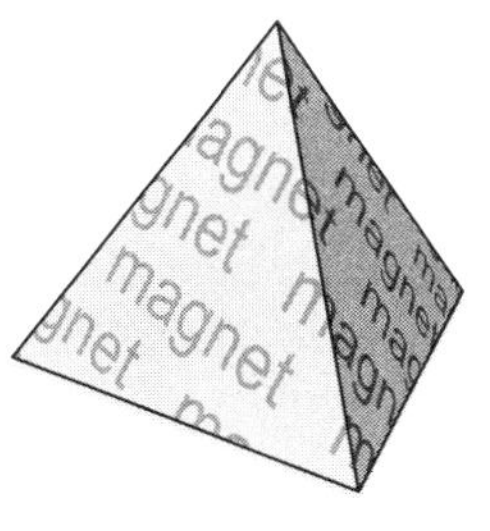

You have a tetrahedron that is a strong magnet. Also, you have a big stock of equal coins. Note that a coin exactly inscribes a face of the tetrahedron.

What is the maximal number of coins that you can affix directly to the surface of the tetrahedron? Coins will be held by the magnet when they touch the tetrahedron's faces (not only its edges or corners) directly with some real (even very small) area of their sides, but never with their rims. Coins can touch each other, but not overlap.

Solution on page 97.

Tetrapaving

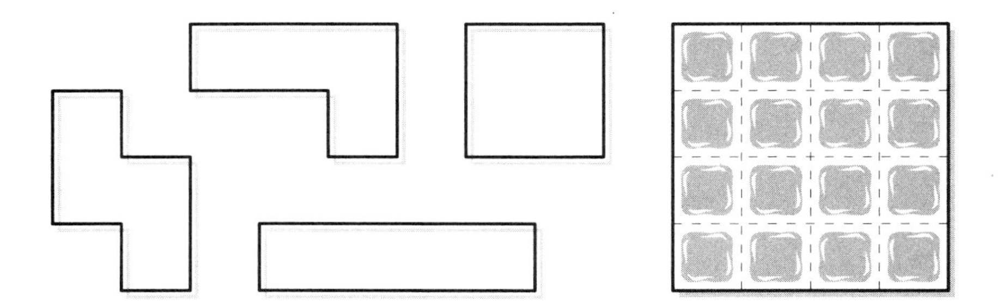

You have four contour (transparent) tetrominoes and the 4 × 4 board. The object is to put all tetrominoes exactly within the board so that it is divided into eight contour dominoes (1 × 2 contour rectangles). You can rotate, flip, and overlap the tetrominoes.

Solution on page 97.

The Yawl

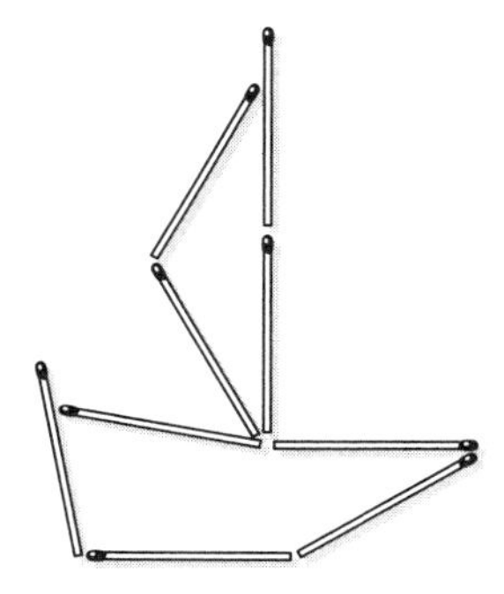

Move five matchsticks so that a yawl of exactly the same shape sails in another direction. Note that in this puzzle loose ends of matchsticks are allowed.

Solution on page 97.

Stairs in the Pyramid

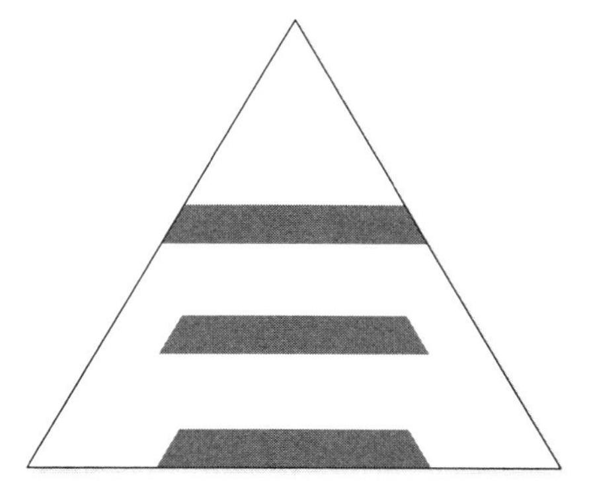

Can you tell (without using any tools) which of the three steps is the widest? Then check your answer with a ruler.

Solution on page 98.

Cubius

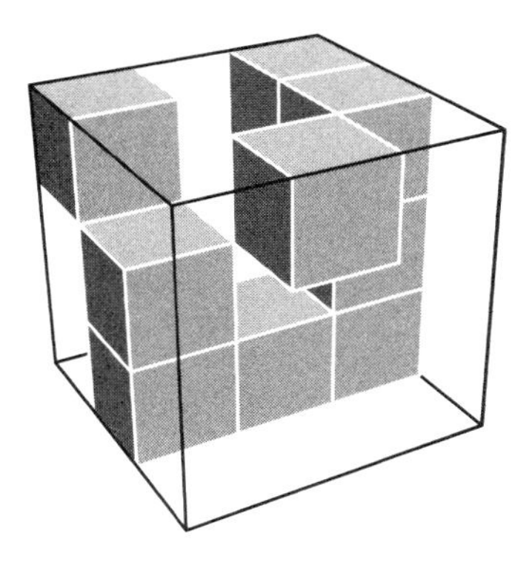

A transparent 3 × 3 × 3 cube contains nine small cubes. What is the message hidden in this composition?

Solution on page 98.

Penta Duo

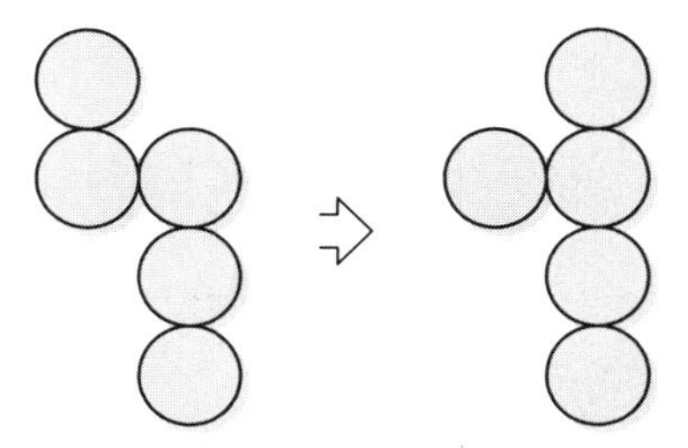

Two pentomino-like shapes are each formed by five identical coins. Observing the "double-touch" rule, reach the goal position shown on the right with the minimal number of single-coin moves.

As a warm-up puzzle, try to solve the backward challenge first, i.e. reach the initial position from the goal one. You can do that in just two single-coin moves, while observing the "double-touch" rule.

Solution on page 99.

The Jigsaw Square Fusion

Divide the two jigsaw pieces into four total parts so that the pieces can be rearranged into a perfect square. The resulting square should have a regular checkered pattern (except for the "ball" parts) like that shown on both jigsaw pieces.

Solution on page 99.

Brick Knights' Swap

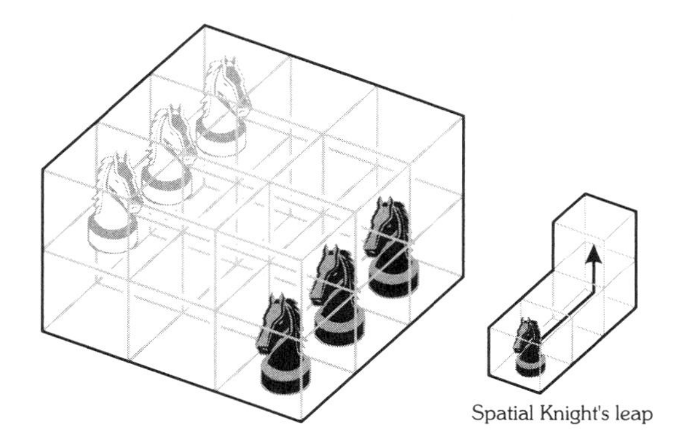

Six chess knights, three white and three black, are placed in the cells of the special 2 × 3 × 3 transparent block as in the big diagram. Now, performing normal knight moves, exchange the white and black knights. Counting a consecutive series of moves by one knight as one move, can you exchange the knights in less than ten moves? A sample of a knight's move inside the block is shown in the small diagram just next to the block.

Solution on page 100.

Solid Chain

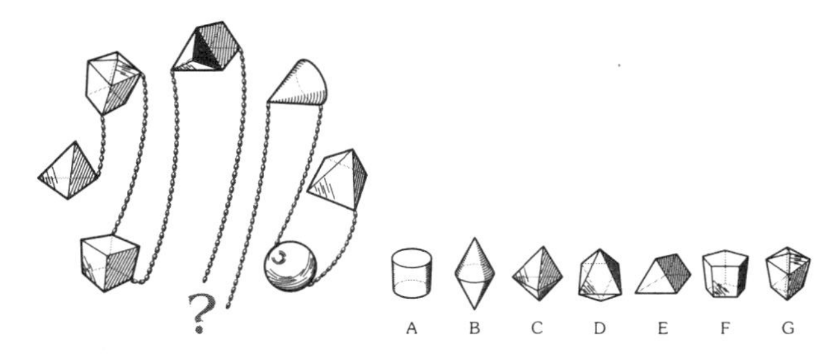

Which one of the seven solids (A–G) should replace the question mark to complete the chain of solids?

Solution on page 100.

The Tomahawk

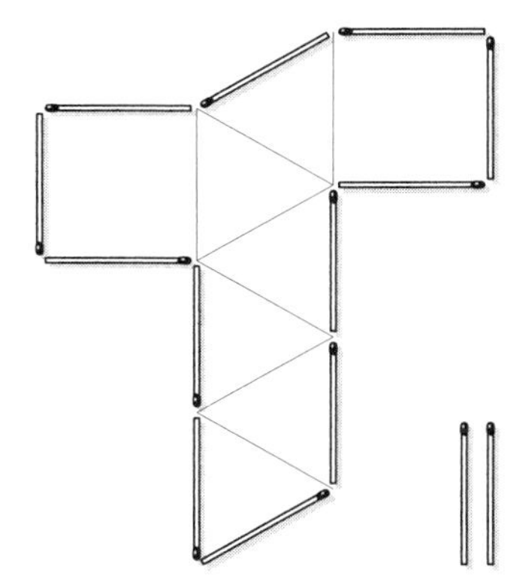

Show how to divide the tomahawk shape into two parts of the same area using two matchsticks.

Solution on page 101.

Delta Cube Score

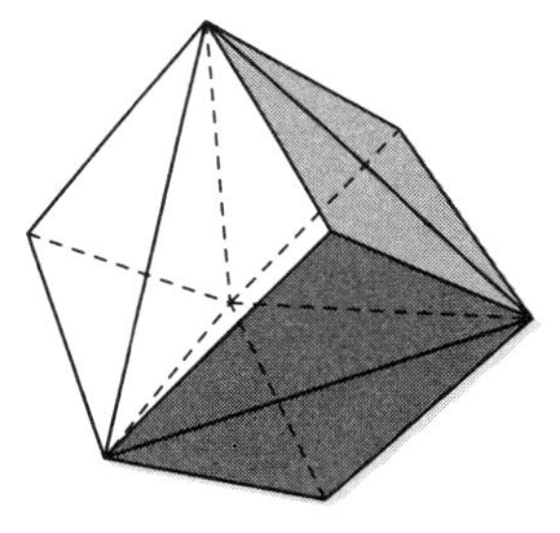

You have a cube with each face divided into two congruent isosceles triangles. How many triangles of any size are on the surface of the cube? The sides of these triangles must be along the lines on the cube (including its edges). Note that you should count triangles assuming that they are made of thin paper and can be wrapped around edges of the cube, having some parts located on different faces of the cube. The triangles cannot be cut or self-overlapped.

Solution on page 101.

Around the Table

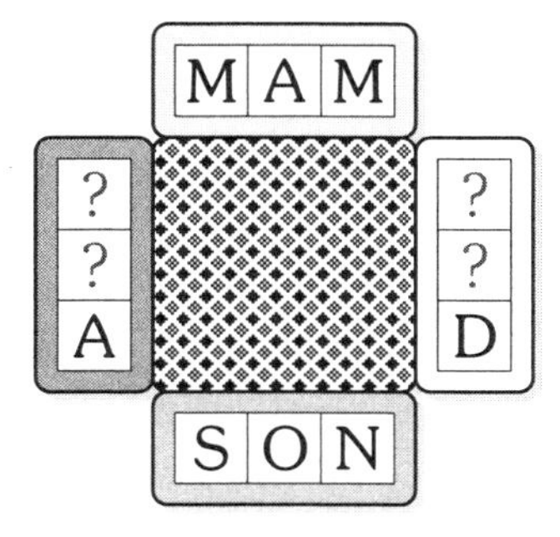

A family is sitting around the table. Two family members' names are already shown. Can you complete the names of the two other relatives, replacing the question marks with the appropriate letters? (Hint: This family has four members only.)

Solution on page 102.

The Three I's

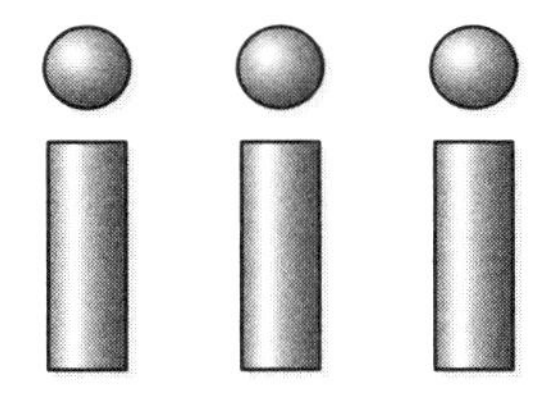

Six pieces are arranged to form three i's. Move the least number of pieces to form exactly:

(a) four i's;
(b) five i's;
(c) six i's.

All i's should have the same shape and size. No unused or overlapped pieces are permitted. How many pieces do you have to move in each case?

Solution on page 102.

Butterfly Differing

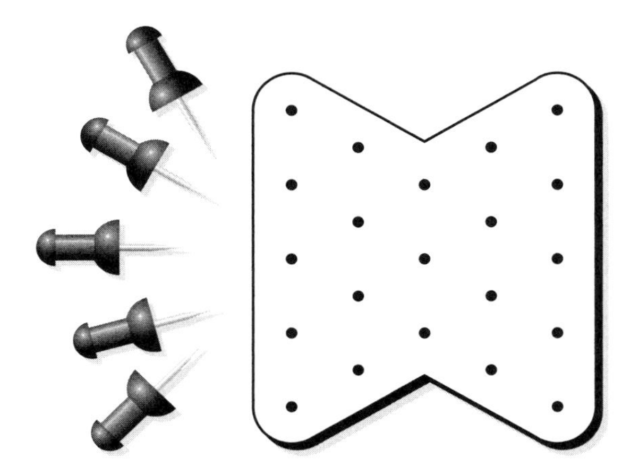

Place five push-pins in five holes in the board so that there are no two pairs of equidistant push-pins.

Solution on page 103.

NumCount

What digit should replace the question mark?

Solution on page 103.

Four Matches & Nautilus

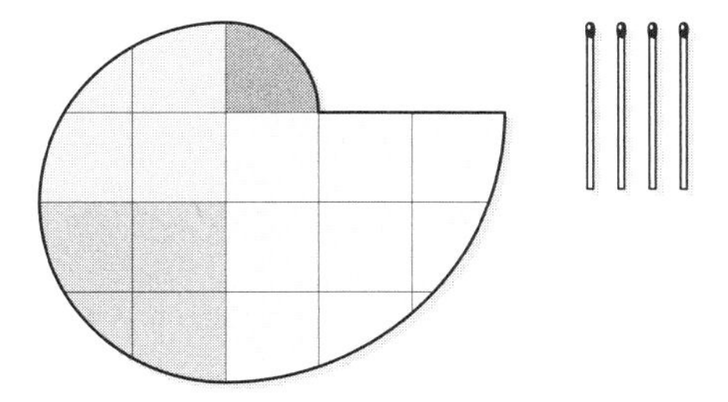

Using four matchsticks, divide the nautilus shape into two parts of the same area. Note that shaded parts of the shape are sectors of three circles with radii 1, 2, and 3 units, respectively. A matchstick is 2 units long.

Solution on page 103.

Mag^{3}netic

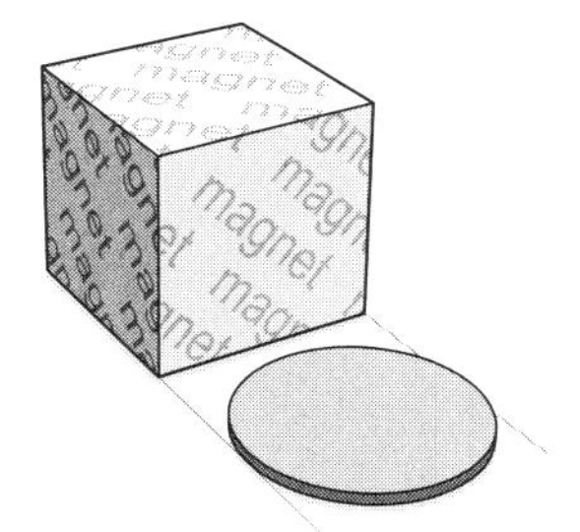

You have a cube that is a strong magnet. Also, you have a big stock of equal coins. Note that a coin's diameter is equal to the cube's edge.

What is the maximal number of coins that you can affix directly to the surface of the cube? Coins will be held when they touch the cube's faces (not only its edges or corners) directly with some real (even very small) area of their sides, but never with their rims. Coins can touch each other, but not overlap.

Solution on page 104.

Coin Upside-Down 2

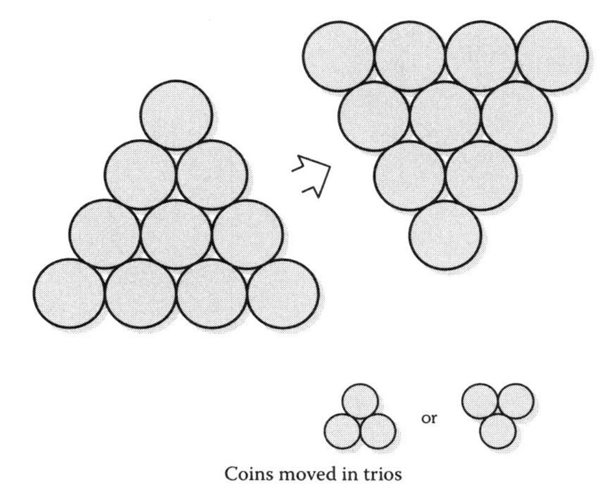

Coins moved in trios

Ten identical coins are arranged into a triangle as on the left. Making only the trio-coin moves, and observing the "double-touch" rule, turn the triangle upside down as shown on the right. Achieve the goal in the least number of moves. A trio-coin move is an orthogonal slide of any three adjacent coins that form a small equilateral triangle as shown in the small diagrams just beneath the goal position.

Solution on page 104.

The Blue Tetrahedron Puzzle

You have four blue (shaded) congruent squares. Put them together to form a tetrahedron with all of its faces colored fully in blue. No folding or bending of squares is allowed. Form the biggest possible fully blue tetrahedron.

Solution on page 105.

Where Is the Solution?

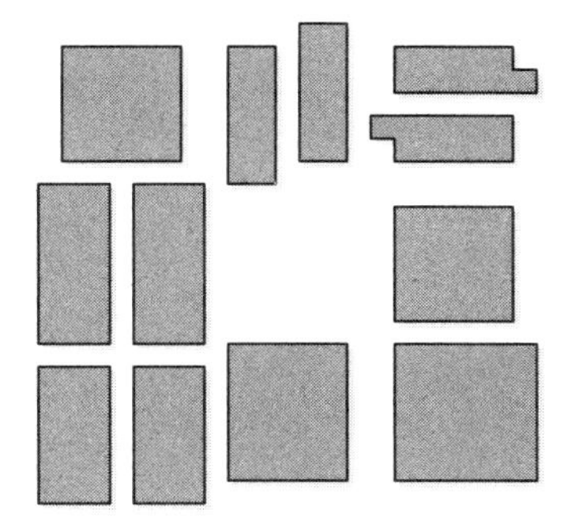

What is the message hidden in the illustration?

Solution on page 105.

Christmas Tree

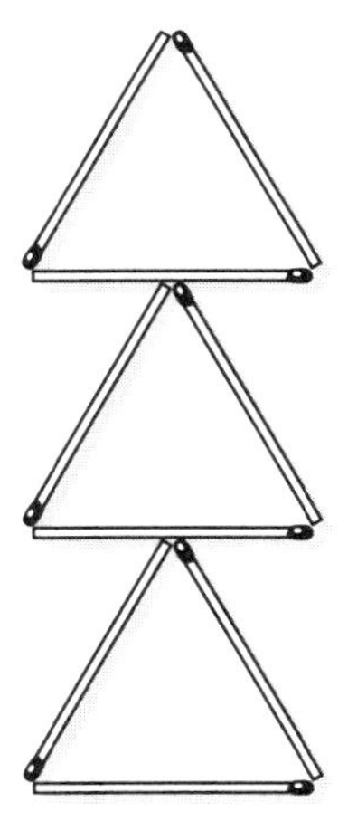

Move three matchsticks so that you get four equilateral triangles on the plane.

Solution on page 105.

Solutions

Introduction to Solutions

Solutions to all the puzzles are presented clearly, sometimes with some further explanations and diagrams provided. In order to show most solutions, the puzzle diagrams are provided again but with additional lines and/or shadings; for example, this technique is used for dissection, dividing, visual, assembling, and dot-connecting challenges.

In the solution diagrams of matchstick puzzles, start and final positions for moved matchsticks are shown as dotted and black matchsticks, respectively. Solutions of sliding-coin puzzles contain sequences of consecutive, move-by-move positions of coins, up to the final positions marked with "F." A moved coin or moved group of coins is always outlined with a bold line, and its destination is always shown with a dotted outline. In the solution diagrams of assembling coin puzzles, the coins in the final shape are shown as they are; if a coin was moved from some start position, that position is shown with a dotted outline, and the moved coin is outlined with a bold line.

For some puzzles more than one solution is possible.

Quadro Block

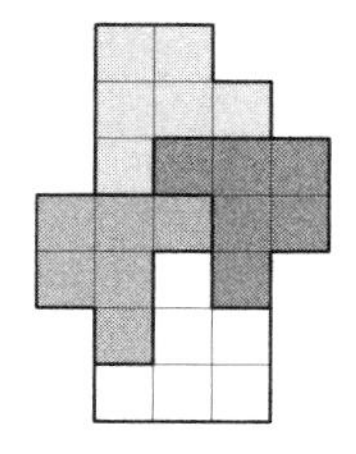

Five Buttons

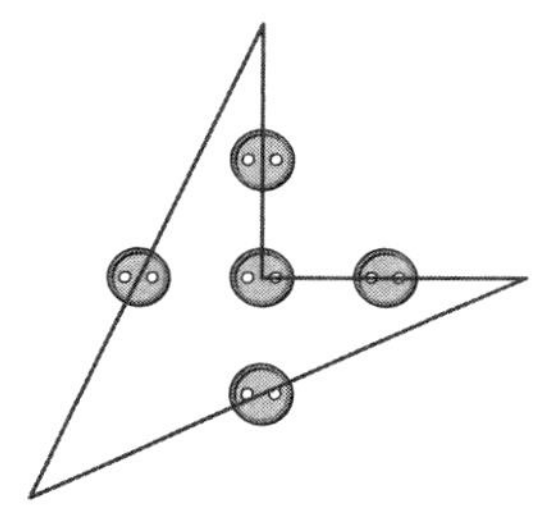

Twenty-Four-Seven

The sequence 24/7/365 translates to 24 hours a day, 7 days a week, and 365 days a year. So, every number deals with some time unit within a larger time span, and that respective span, in turn, is the basic unit for the next number in the sequence. Number 365 breaks the sequence because it deals with days as units as does number 7. In order to make the sequence logically true, the last number should have weeks as units within the bigger time span of a year. Thus, it should be "24/7/52," or in other words, "24 hours a day, 7 days a week, and 52 weeks a year."

Out of the Y

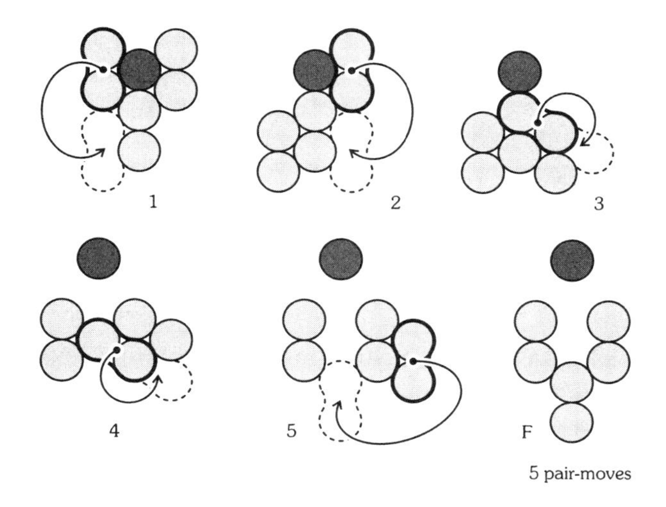

Broken Square

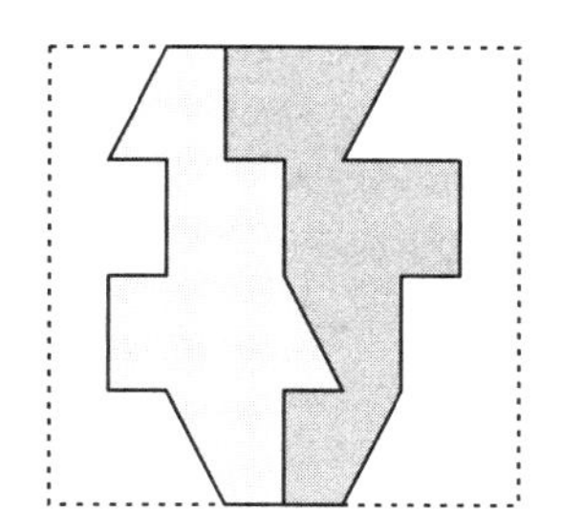

Why Tee Tee

The YTT sequence is the first letters of the words Yesterday, Today, and Tomorrow, respectively. In a week there is only one sequence of days in which the first letters of the days can replace the YTT letters properly. These are Friday, Saturday, and Sunday, or FSS. Thus, the puzzle was aired on Thursday.

Elliptic Proportions

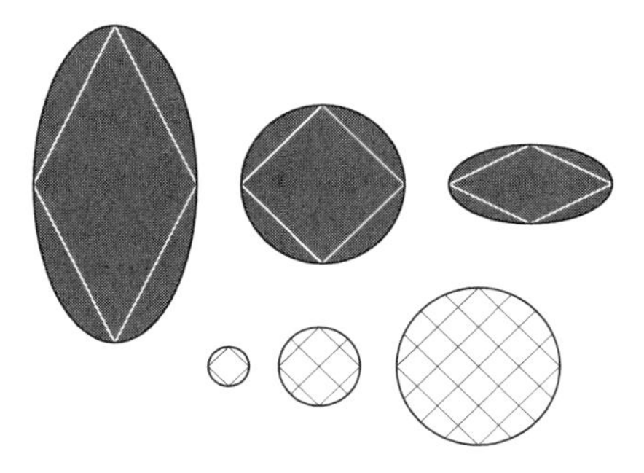

Both ellipses with inscribed rhombuses are obtained from a circle with an inscribed square, as shown in the uppermost diagram. Thus the ratio of their areas is 4:2:1. The areas of the small circle, medium circle, and big circle are in proportion 1:4:16. This leads to the result that the ratio of the shaded areas to the unshaded areas of the shape is 5:11.

Netting

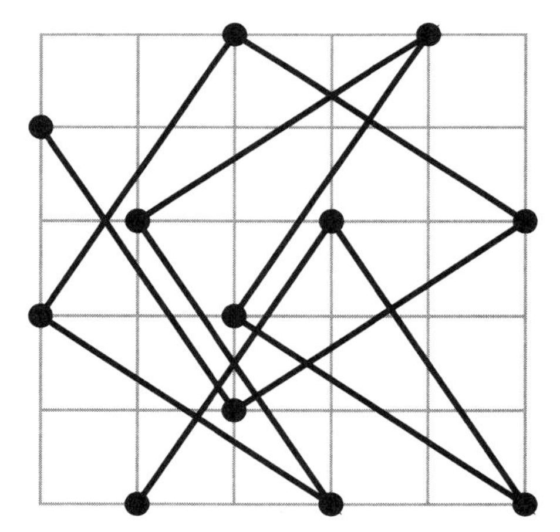

In the Right Triangle

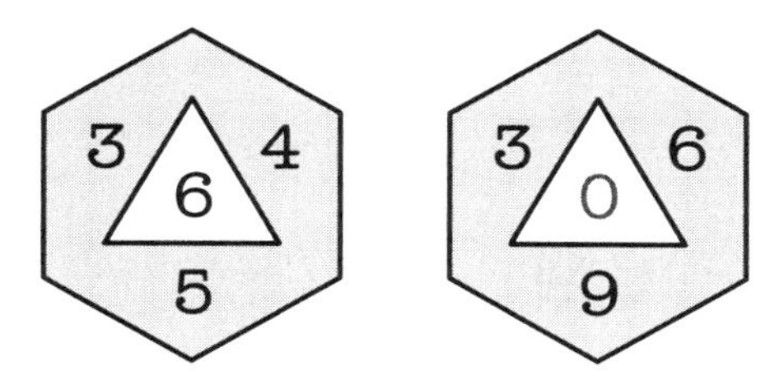

For both triangles the number by each side represents the length of the side, while the number inside the triangle represents the area of the triangle with such sides. For the first triangle the sides are 3, 4, and 5; this is the so-called Egyptian triangle. Its area is $(3 \times 4)/2 = 6$. For the second triangle the sides are 3, 6, and 9; this makes a singular triangle; its area equals 0. Thus, the one-digit number that should replace the question mark in the second triangle is 0.

Checkered Challenge 58

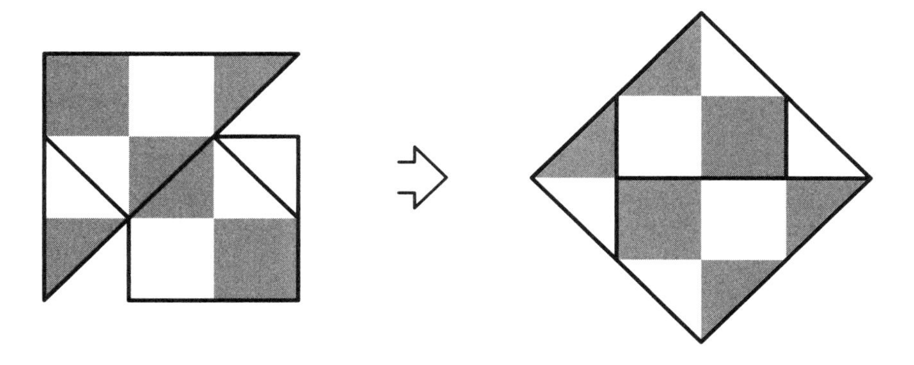

Two Unicursal Bricks

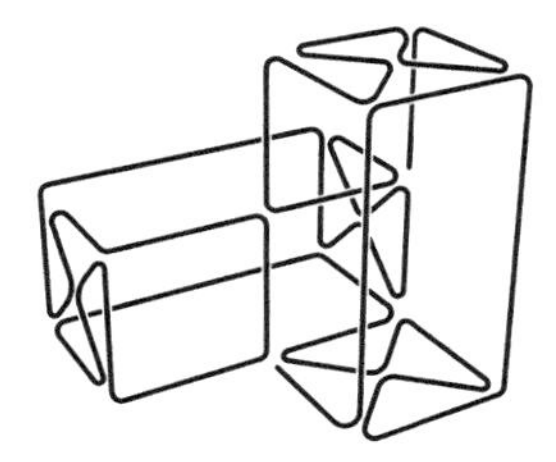

Profiles: The Solids

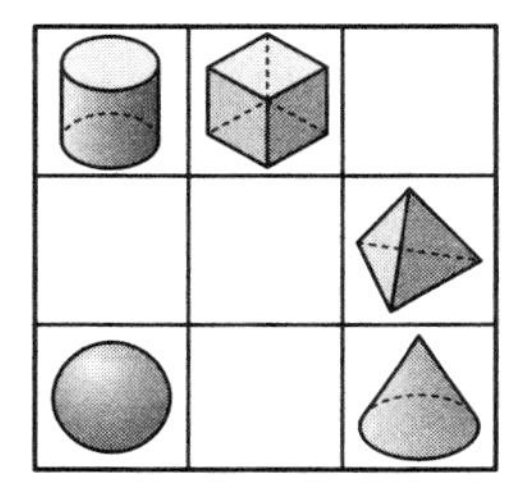

The Wrench Dividing

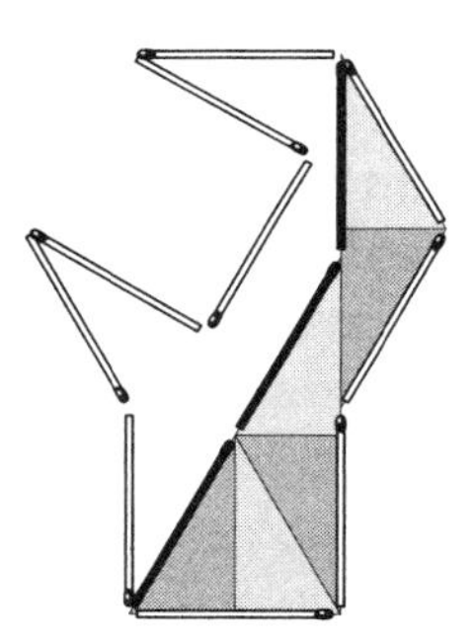

The entire area of the wrench is equal to the area of a regular hexagon of side length 1 (the square "handle" fits in the open end of the "head"). Thus, it equally divides into six equilateral triangles or 12 right triangles. The right part of the divided wrench contains exactly six right triangles, which is exactly half the wrench's area, thus the wrench is divided into two parts of the same area.

In the Same Plane

The Triangle Quest

The maximum number of equilateral triangle outlines that can be formed on the plane from the three congruent equilateral triangles is seven. The basic solution pattern for this is shown in the illustration.

A Smooth Way through the Maze

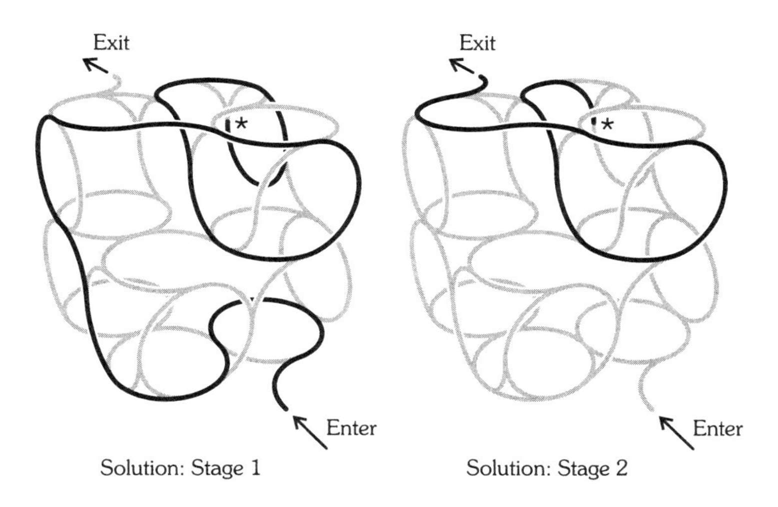

Solution: Stage 1

Solution: Stage 2

Elastic Trios

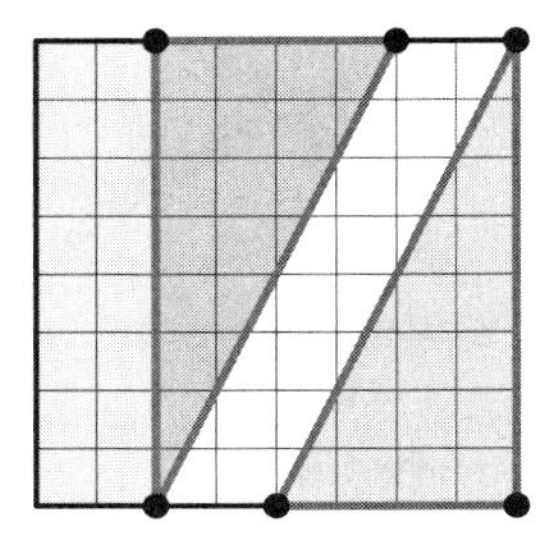

The Tube Colors

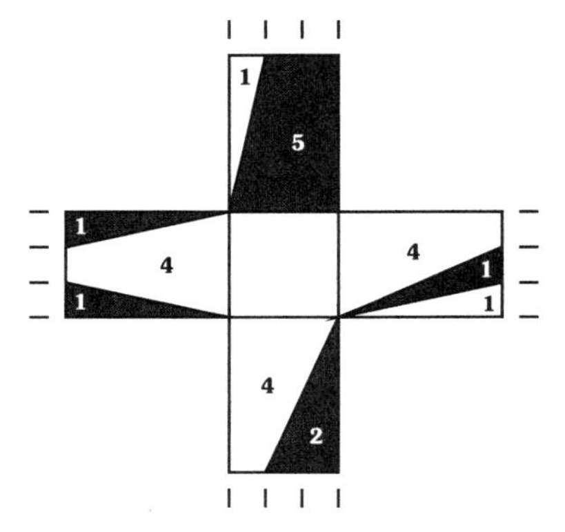

The unfolded tube with the real pattern of its internal walls is shown. Thus, the ratio of the black parts to the light parts of the pattern is 5:7, and more of the pattern is light.

The Air Bubble Challenge

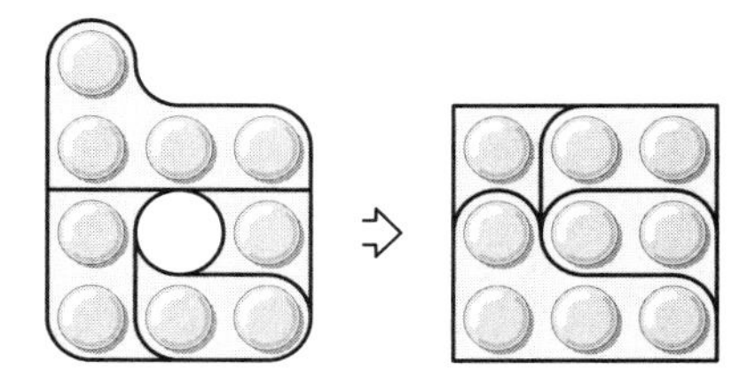

Three Fragments

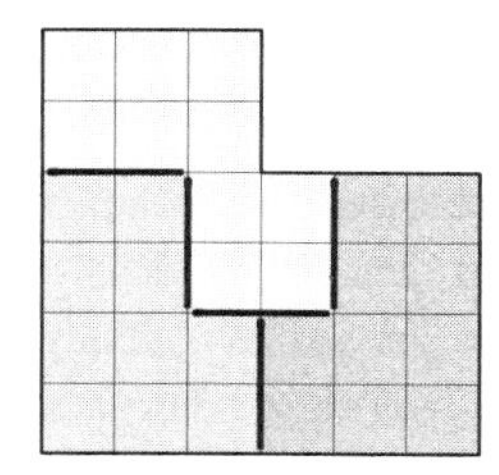

An Odd Field

The three characters provided in the odd application field are, actually, arrows pointing Up, Right, and Left, respectively. The first letters of these three words are "U", "R", and "L", which together are the common three-letter acronym "URL." In other words, the applicant filled in this field with the URL address of his website. Of course, it can be interpreted that the arrows point to the North, East, and West, resulting in the word NEW. But, since the hint states that the field was not NEW to the applicant, "URL" is the only relevant meaning of the code.

Finding the Solution

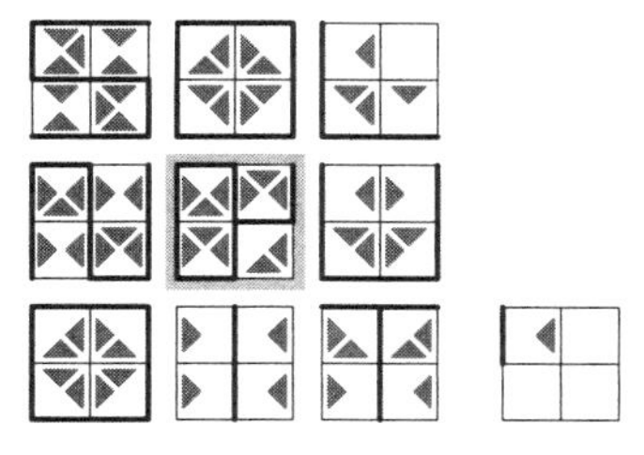

Every triangle in each small square is pointing at a line segment on the opposite side of the square, as shown in the small diagram. In some cases a line segment is common to two adjacent squares, so two triangles are pointing at it. As a result we can write eight letters that, starting from the top left one and going clockwise, spell SOLUTION. The pattern in the middle square gives the shape of a key.

Pentatriangles

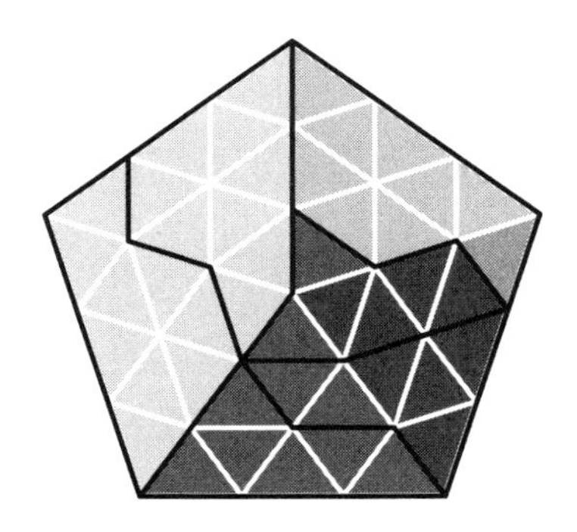

The Drop

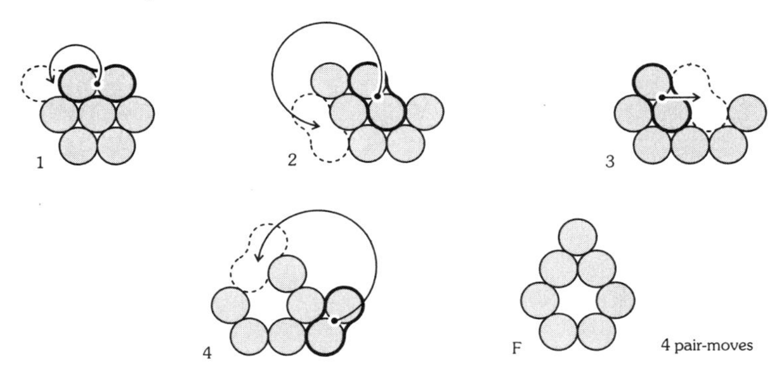

Triangle & Two Matchsticks

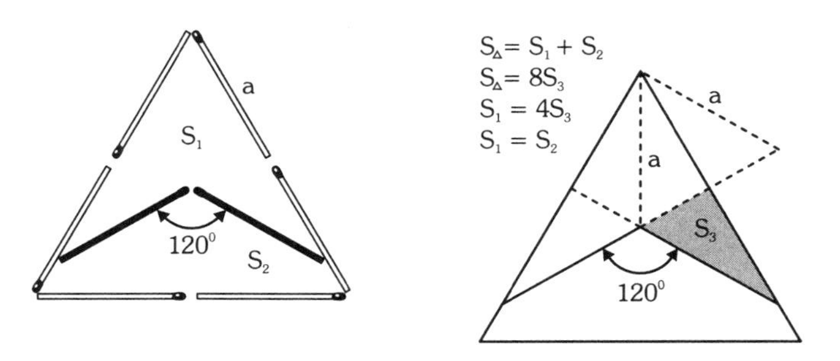

Easy L-Packing

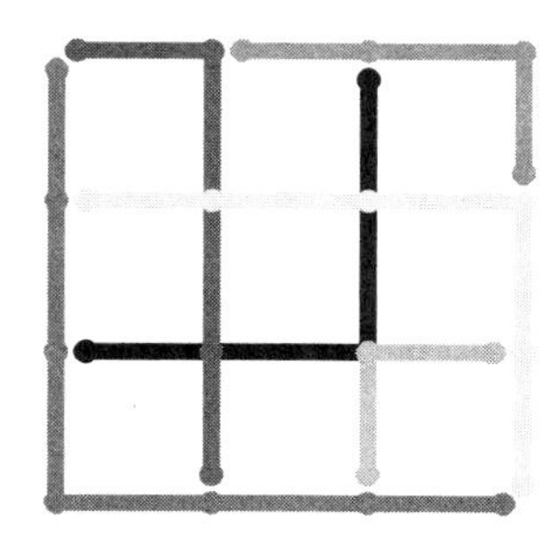

Golden Budget

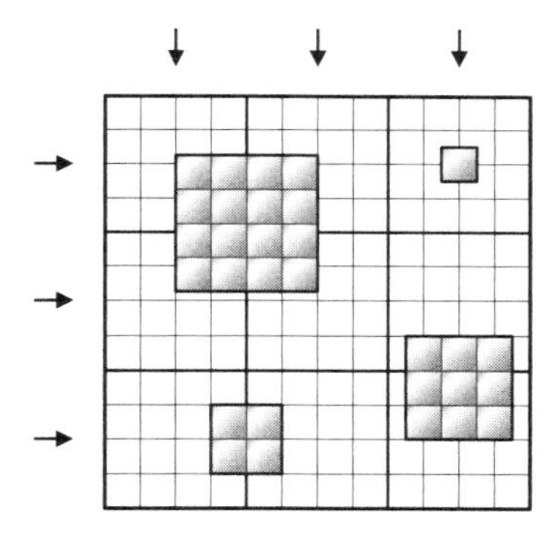

The Shape-Color Connection

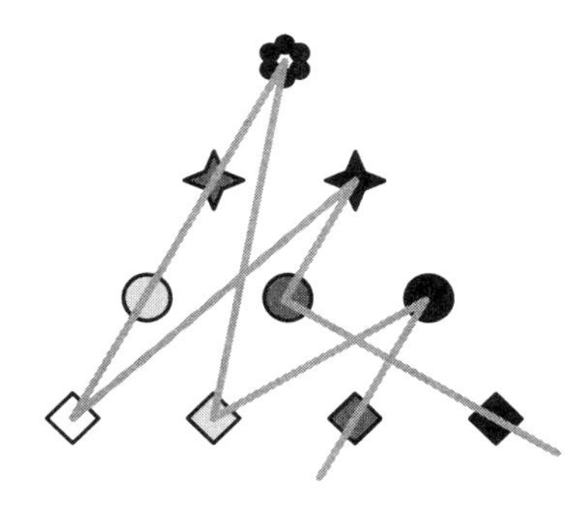

Separate the Shapes

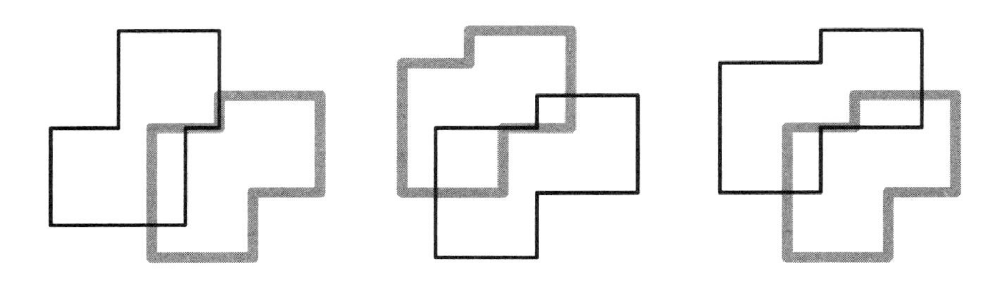

S-days and T-days

The days that start with "S" are Saturdays and Sundays. The days that start with "T" are Tuesdays and Thursdays. There are only two pairs of consecutive months that can have an equal number of S-days and T-days (9 and 9) and the same total numbers of these days in each of them (18). These are December and January, and July and August (consecutive months with 31 days each). Also, it can easily be seen that for the special day counts to be true in either pair, the second month must begin on Sunday. So, since the two friends' conversation happened on the first day of a month, well after all winter holidays, they are talking about the July–August pair, and their conversation took place on Sunday, August 1st.

The Puzzling Cross

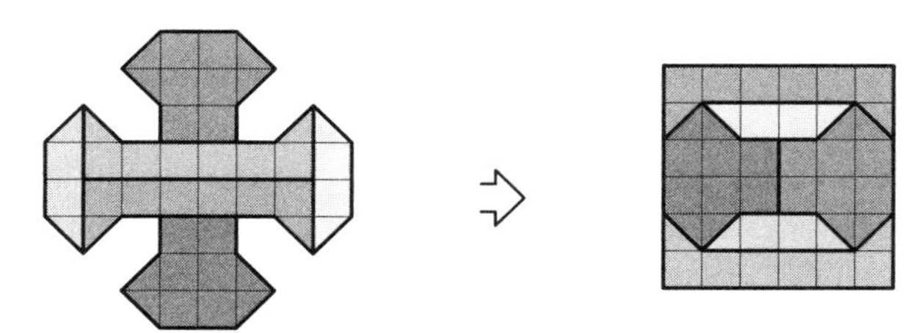

Round-Up

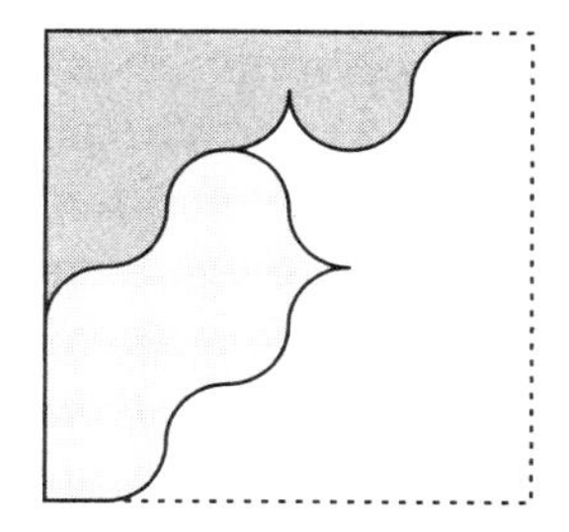

The Square-Triangle Couple

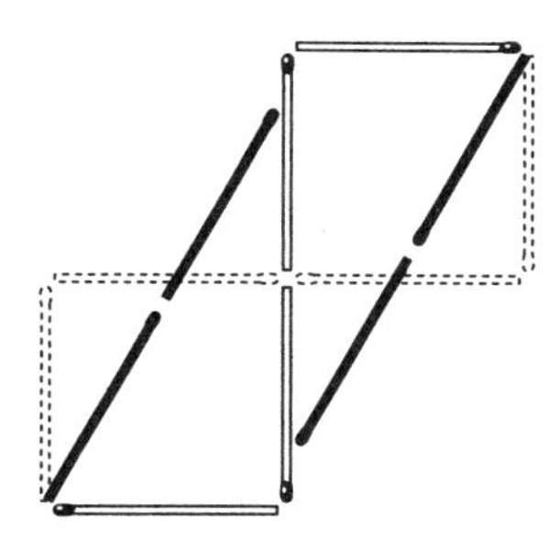

The Die Stack

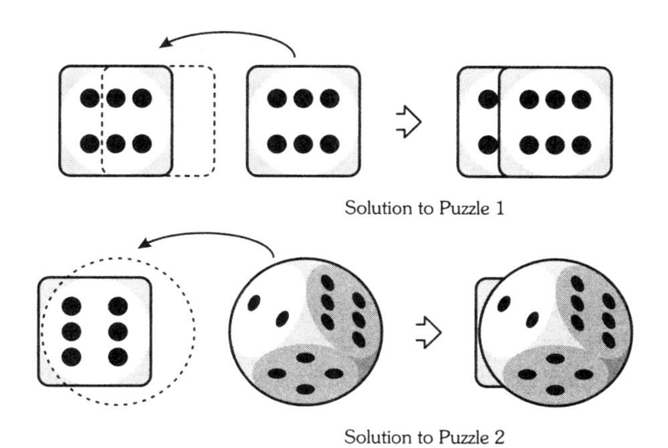

Solution to Puzzle 1

Solution to Puzzle 2

T-Unicursal

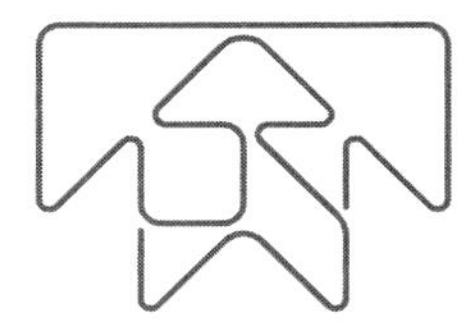

The Square Quest

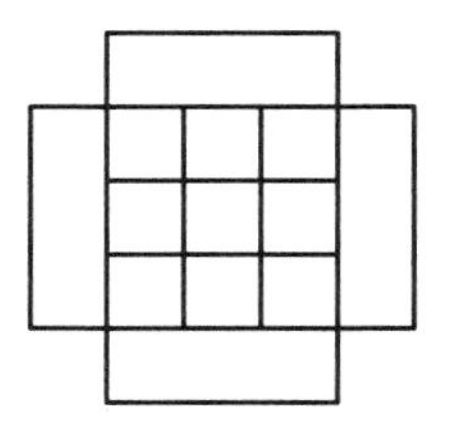

The maximum number of square outlines that can be produced on the plane from the four identical square frames is 18.

Signs & Symbols

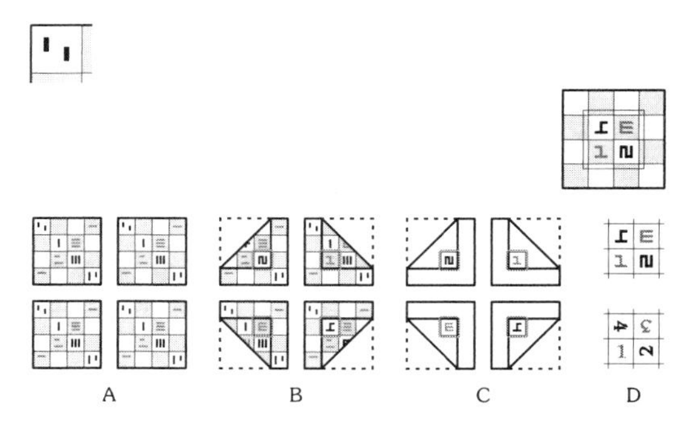

The symbol on the top left should be placed in the top left cell of the checkered 4 × 4 paper square. If you fold the square in the four different ways as shown in the diagrams A through D, you will see four different stylized digits—1, 2, 3, and 4—which appear when two cells of the square overlap each other.

1-2-3 Transforming Puzzle

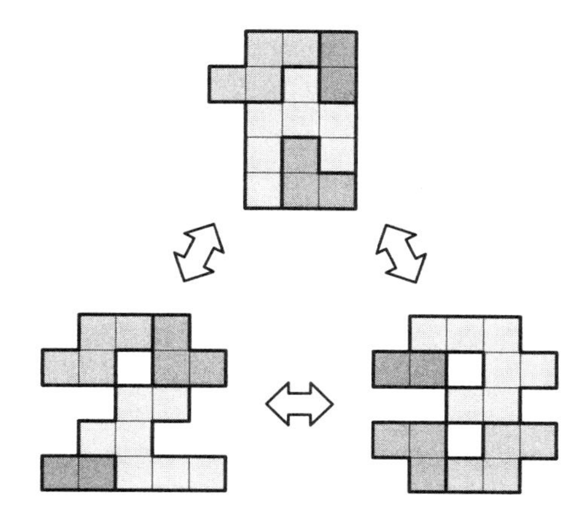

The Viewpoint

Segment A will get longer, while segment B will get shorter.

Play with an Erasure

Some possible pairs are listed:

dent => cent
die => lie
nest => rest
nope => rope
tend => lend
took => tool

Trapezoid Contours

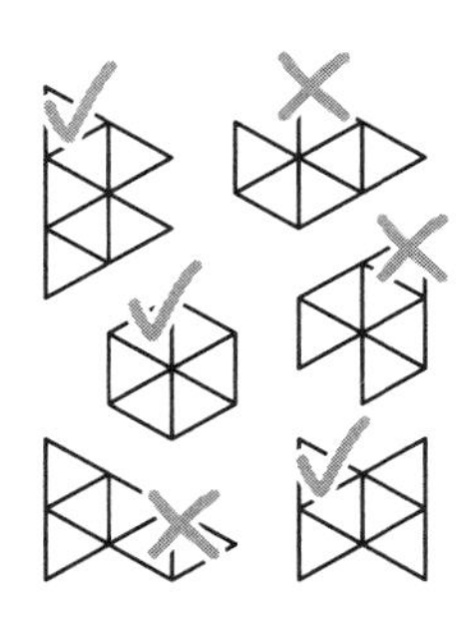

Add a Row

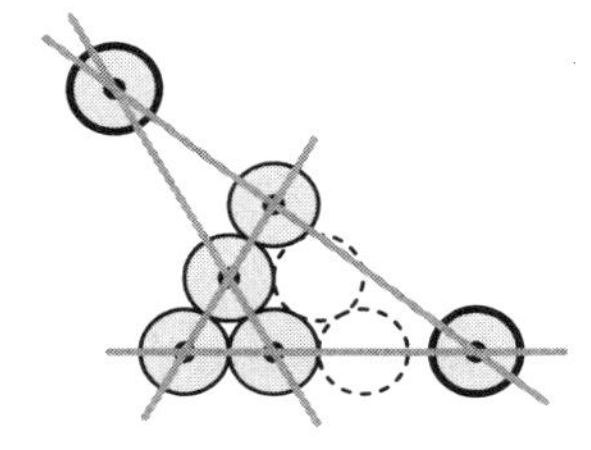

The M-shaped Count

In the grid you can find 60 mitre shapes of four different sizes; they are shown in the four grids.

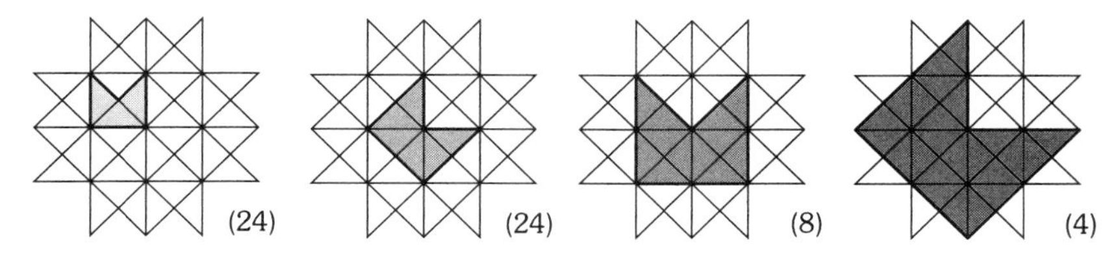

Note that the shapes have different orientations in the grid. The numbers in parentheses show the total numbers of copies of each shaded M-shape hidden in the grid, for a total of 60 M-shapes.

No Magic

The digit to replace the question mark is 7. The several digits presented in the puzzle are part of a traditional 3 × 3 magic square.

What Dice It Matter?

Hexa Differing

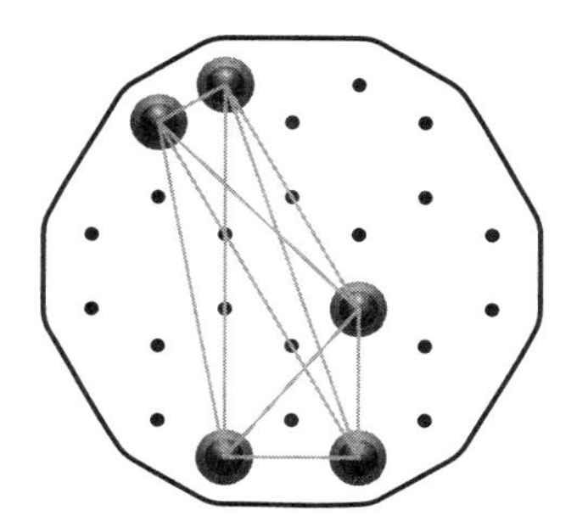

The Heart of the Match

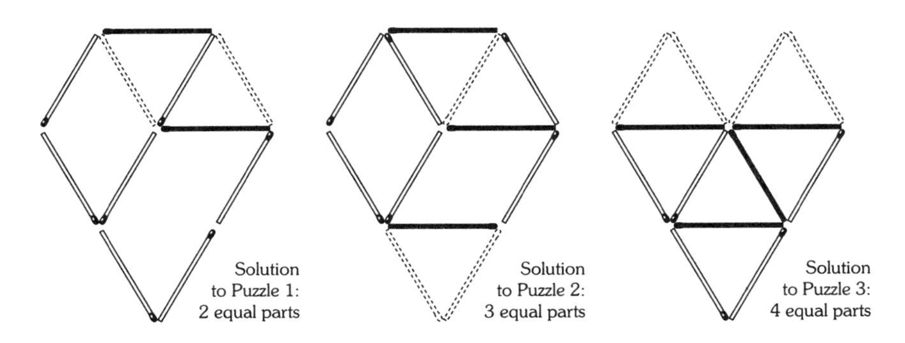

DigitCount

87355

The digit 5 should replace the question mark. The progression is that each digit is the number of segments used to form the previous digit.

The Book Staircase

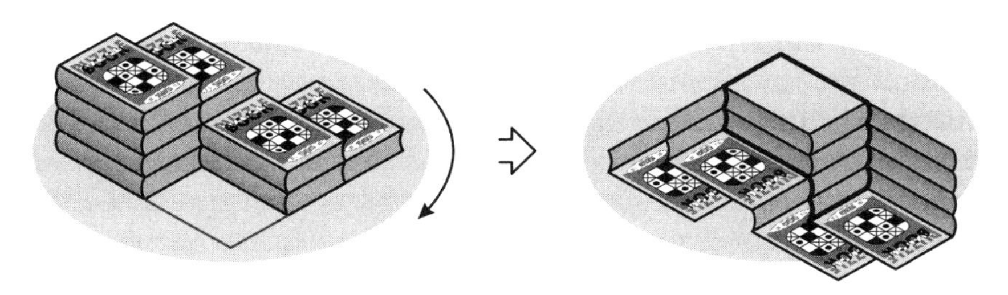

Draw two straight lines as in the illustration on the left. Then turn the page 180°. Now you can see several additional books; they are outlined with the bold line.

Two Different Triangles

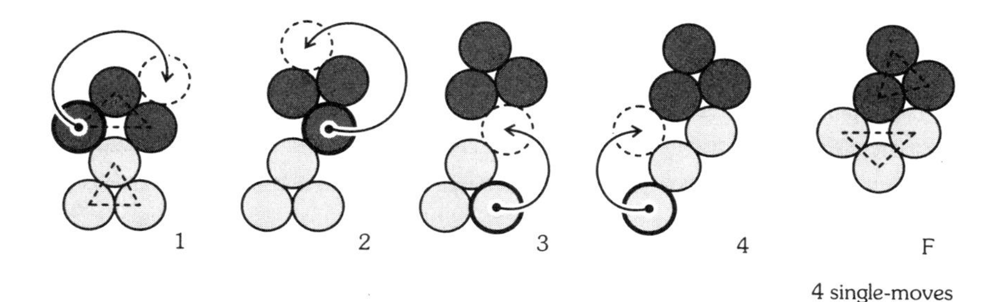

Inside the Grid

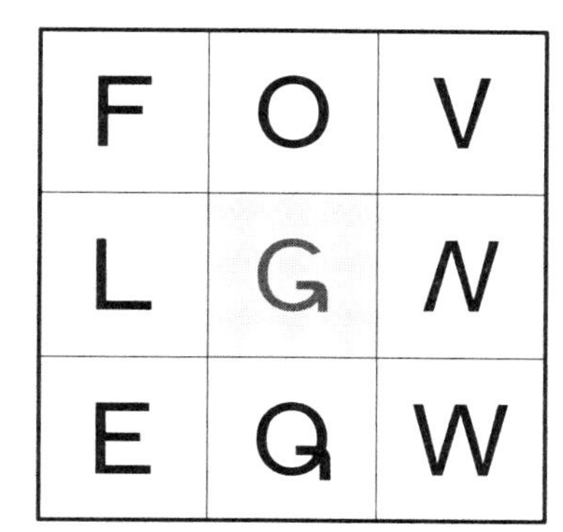

In the first column, when F and L are superimposed, E is produced. In the third column, when V and N are superimposed, W is produced. In the second column, G is missing in order to produce Q when superimposed with O.

Drop & Match

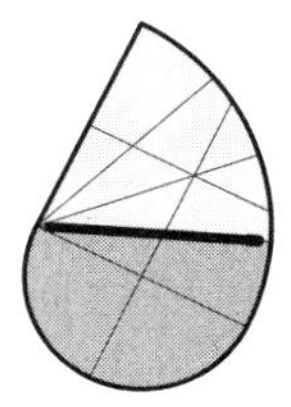

Mag²netic

You can affix 14 coins to the square magnet (seven coins to each side) as shown in the diagram.

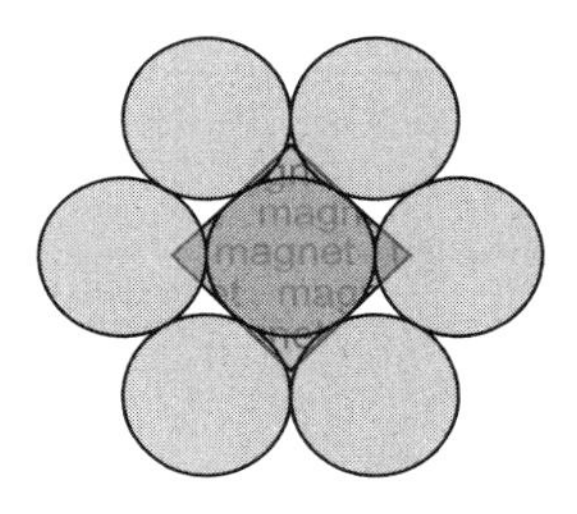

The Antique Ring

Each pair of patterns in the opposite openings always gives the same pattern. The missing wire element is highlighted.

The Factory Block Puzzle

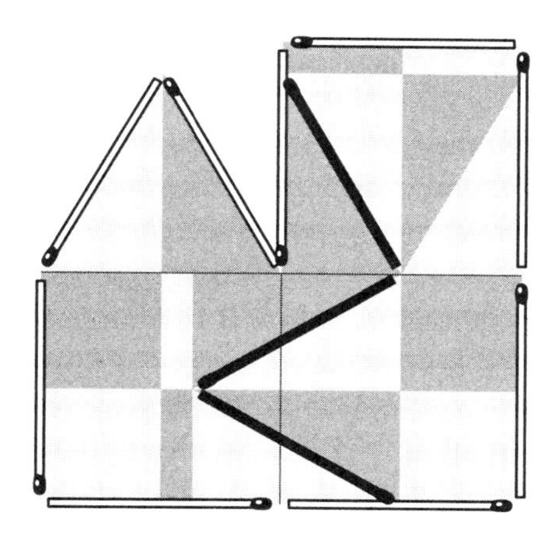

Puzzle Card

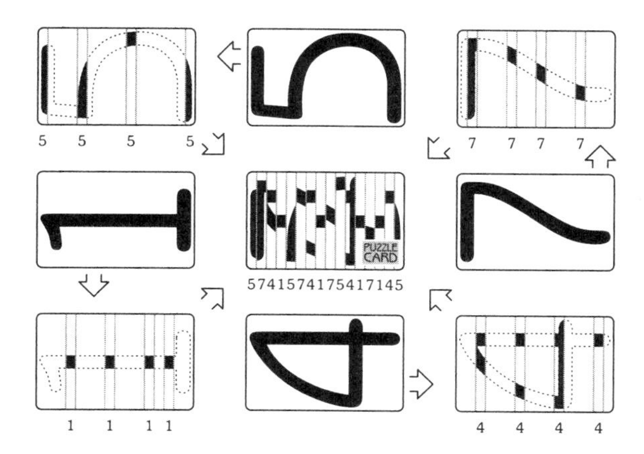

The pattern on the Puzzle Card is assembled from 16, single strips taken from the four separate "digital" cards with big digits 1, 4, 5, and 7 and placed on the Puzzle Card in their respective places. Thus, each digit in the 16-digit number represents the strip and digit of the respective "digital" card.

Coin Upside-Down

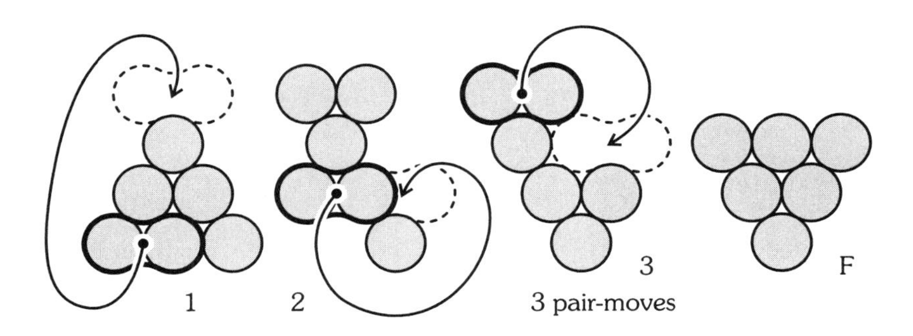

The Puzzle Infinity

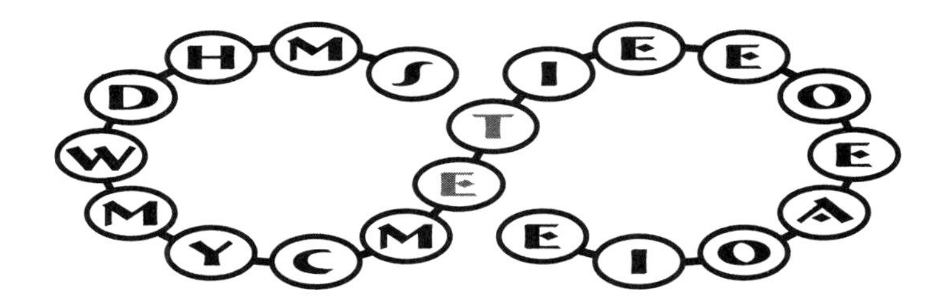

Let's go simultaneously from both ends of the infinity line taking two letters at every step (left + right). We can see the next pairs: SE, MI, HO, DA, WE, MO, YE, CE, MI. These pairs are the first letters in the following words that mean different time periods starting from the shortest: SECOND, MINUTE, HOUR, DAY, WEEK, MONTH, YEAR, CENTURY, MILLENNIUM , ... The final one can be ETERNITY, so the two letters replacing the question marks are E and T.

The Caravel

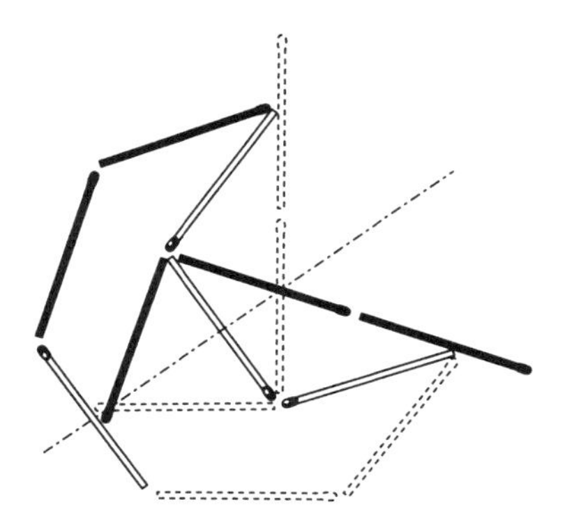

Twin Cubism

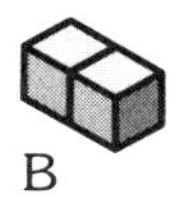

All three cube-twins in the diagram are the same but shown from different points of view. Thus, the answer is cube-twins B.

Three L's to a T

Equal Perimeters

What's There in the Square?

x	1	2	3
1	1	2	3
2	2	4	6
3	3	6	9

The square grid is a multiplication table. Each number is the product of the respective row and column numbers. The missing number is 9.

G-Knights Exchange

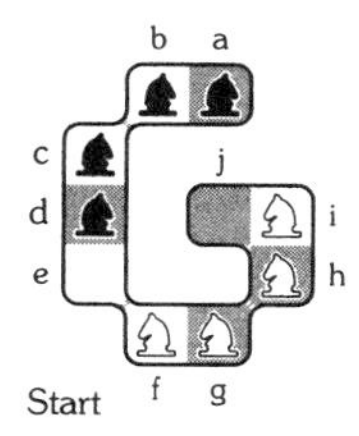

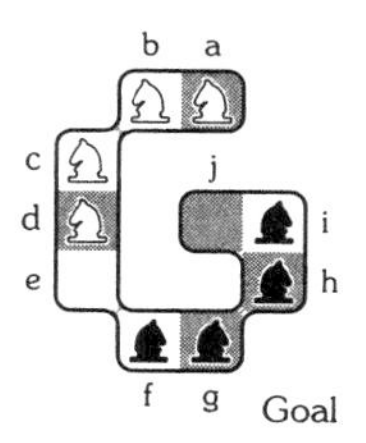

To exchange the knights you will need the 11 moves listed:

1.	b-j-e	7.	a-c-j
2.	d-b-j	8.	i-a-c
3.	f-d-b	9.	g-i-a
4.	h-f-d	10.	e-g-i
5.	j-f-h	11.	j-e-g
6.	c-j-f		

Rectangle Differing

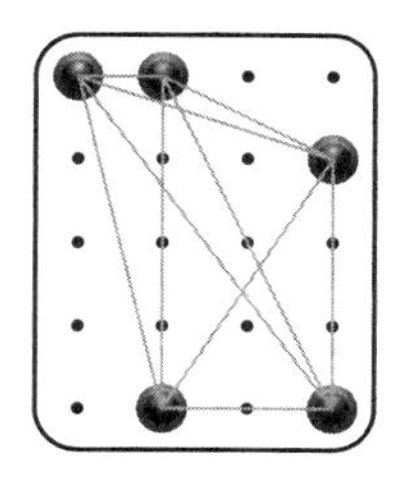

Four More Triangles

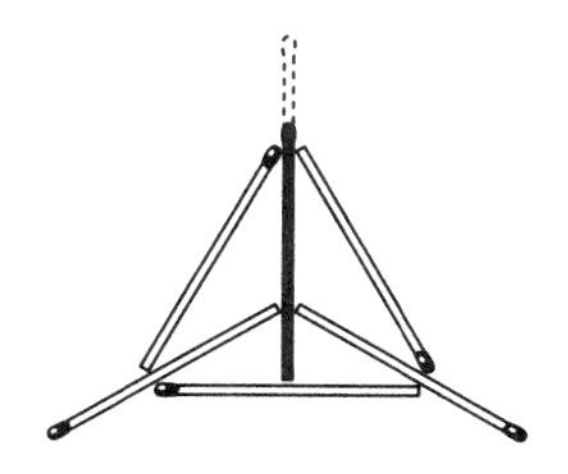

The Four Snakes Puzzle

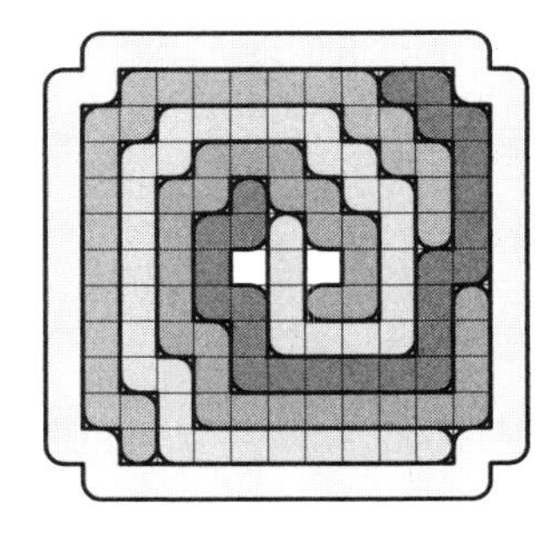

Black or White in Cube

The ratio of all the black parts to all the white parts of the cube's surface is 1:1.

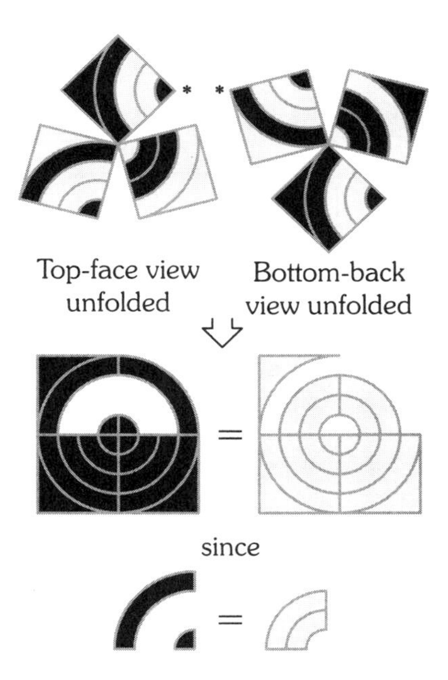

Pentomino Switch

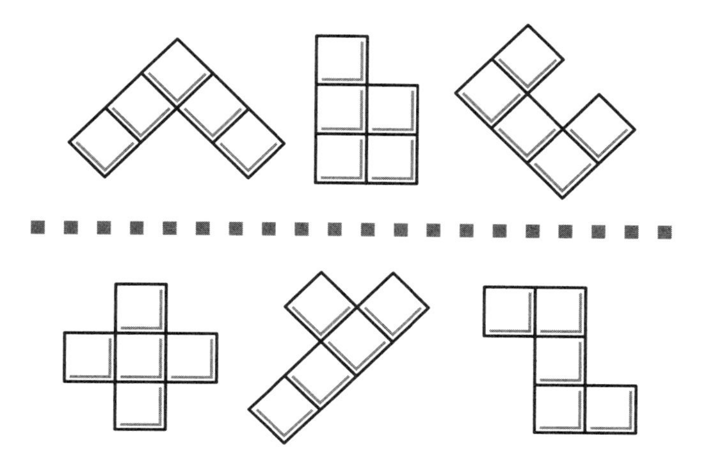

The two pentominoes that should be exchanged are the upper-left and the lower-right ones. When they are exchanged and placed as shown, the original pattern is restored: the upper row of pentominoes is A, B, and C; the middle 20 dots are the next 20 undisclosed letters of the English alphabet; and the lower row of pentominoes is X, Y, and Z.

The Ancient Pyramid Puzzle

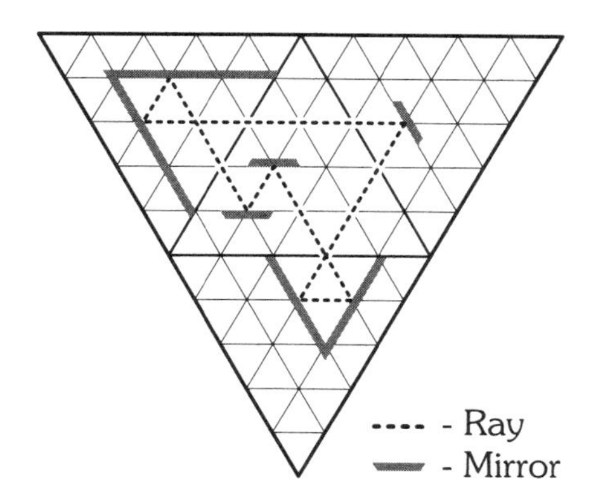

Broken Watch

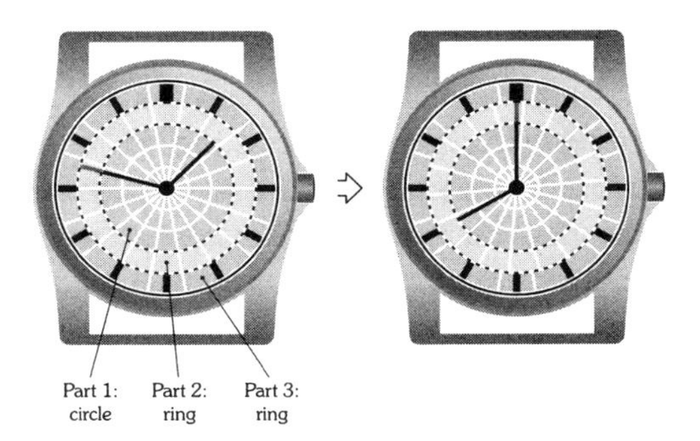

In the solution, Part 1 is rotated 45 degrees counterclockwise, Part 2 is rotated 75 degrees clockwise, and Part 3 stays as it is.

Coins Apart

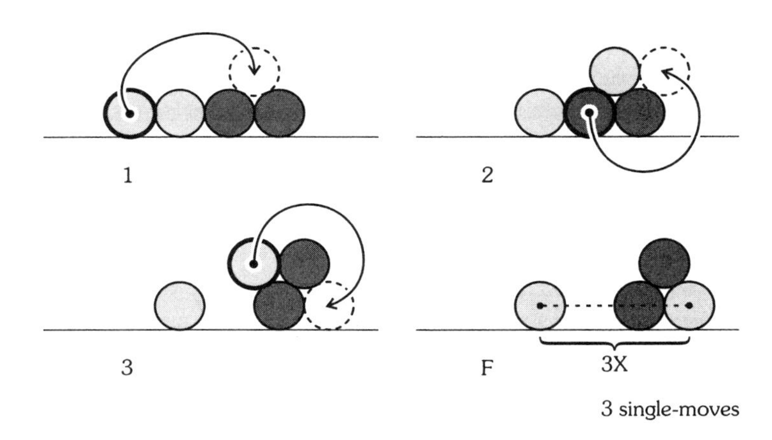

The Deer Puzzle

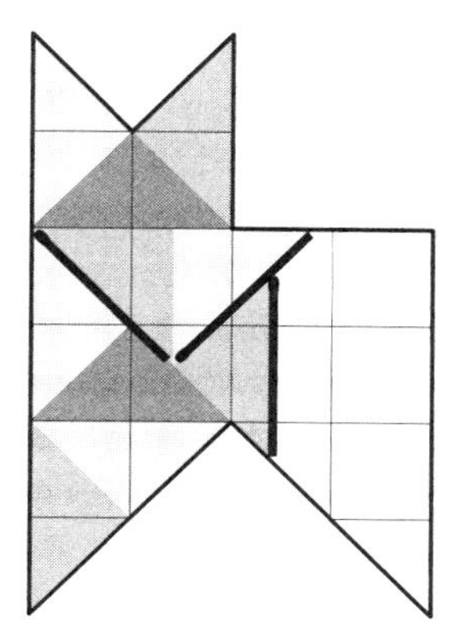

The total area of the deer is equal to 15 small square cells of the grid, or 15 right isosceles triangles of the same area. Each of the two shaded parts of the divided deer contains exactly five right isosceles triangles, which is exactly one third of the deer's area, thus the deer is divided into three parts of the same area.

Cheap Victory

The only four capital letters in the English alphabet that can be constructed from only two equal straight planks joined together are those shown. The letter **V** completes this set.

Down the Street or… Finding the House

House #5 of Horizontal Street is on block C. The house numbers in each block are shown in the illustration.

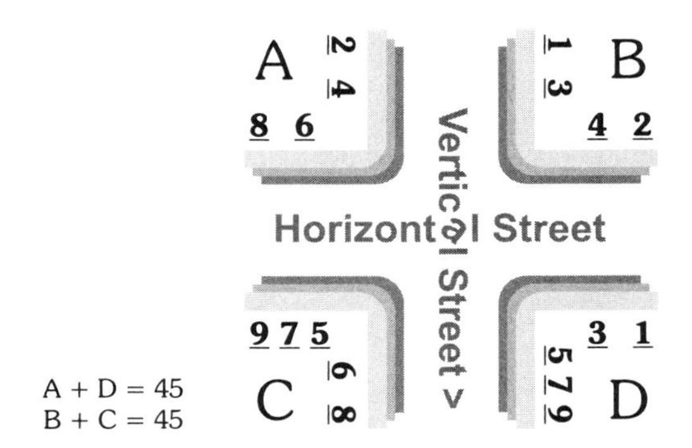

97 Question 5

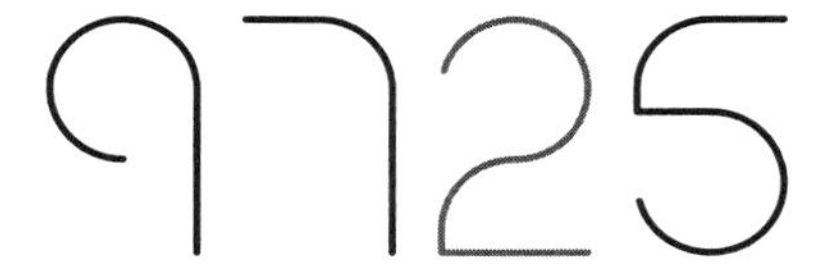

A 2 should replace the question mark, completing this pattern: 9 – 7 = 2; 7 – 2 = 5.

Triangles & Digits

Coin Invert

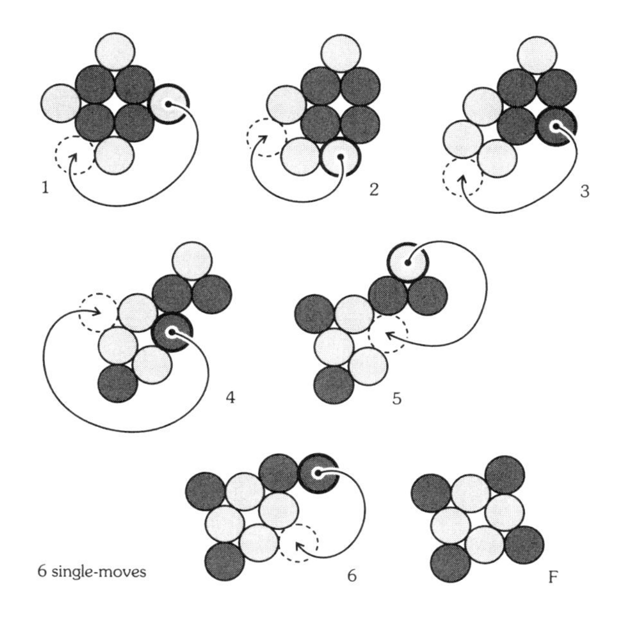

The Legendary Town

The pattern of the town's street hides the following letters—S, U, N, C, I, T, and Y—as shown in the diagrams below. This spells the name of the famous Utopian city, the Sun-City, or the City of the Sun, described by the Italian philosopher Tommaso Campanella in 1602.

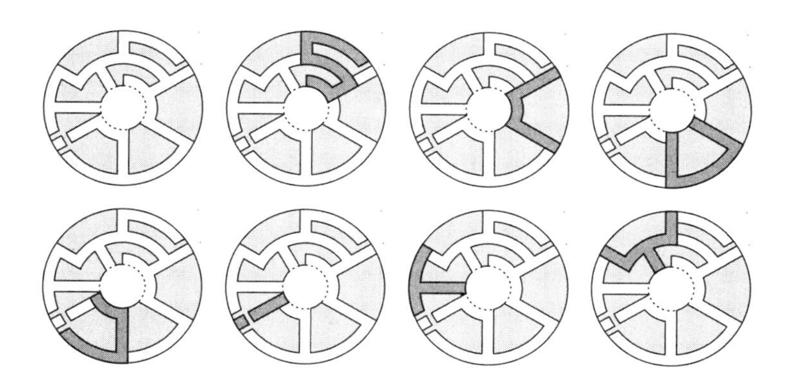

Cubism

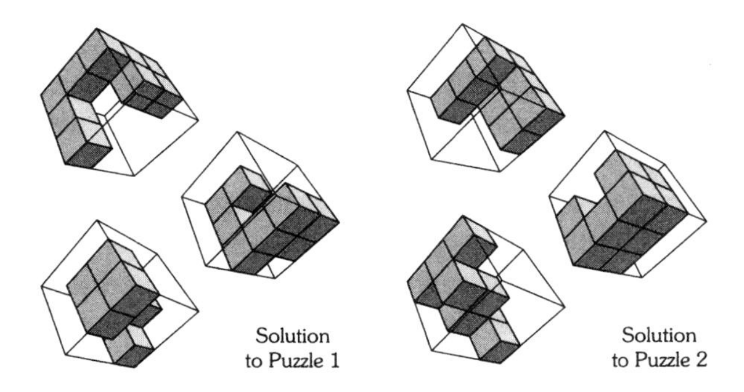

The Matchstick Needle

Shore Connecting

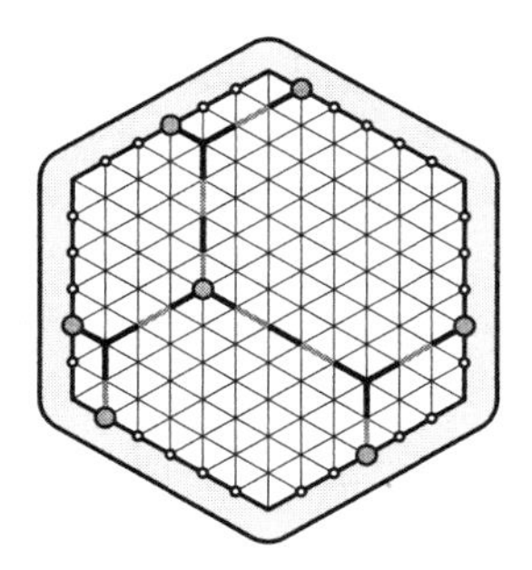

Cube Differing

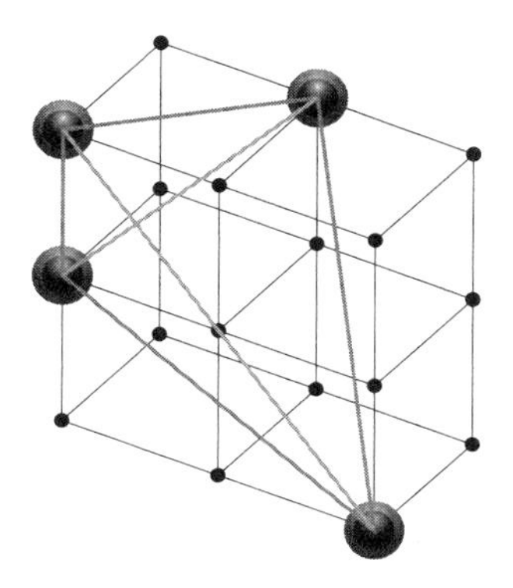

The Right-Angle Framework

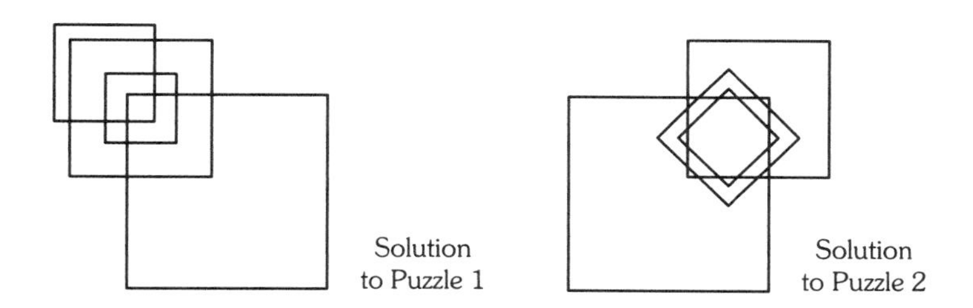

Puzzle 1. The maximum number of perfect square outlines of any size that can be produced on the plane from the four perfect square frames of decreasing sizes is 13.

Puzzle 2. The maximum number of right isosceles triangle outlines of any size that can be produced on the plane from the four perfect square frames of decreasing sizes is 16.

The VHS Tricky Packing

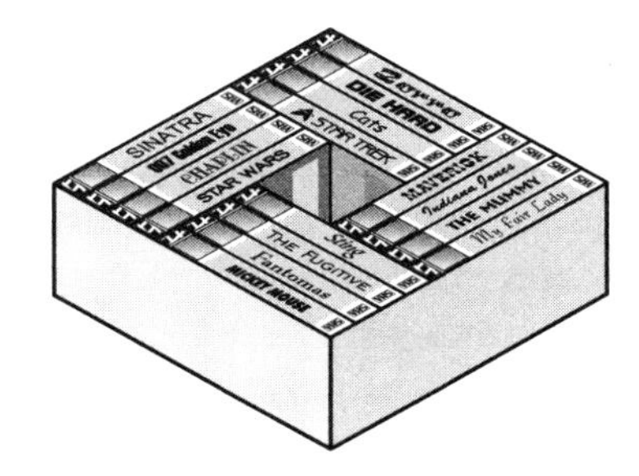

Coin Triangle Theorem

Each single move "turns" the triangle upside down. Thus, to get the final triangle turned with its top down, you always need an odd number of single flipping moves.

Each single move flips a coin from one side to another (heads up to tails up, and vice versa). Thus, to get all coins in the final position with all heads (A's) up, you always need an even number of single flipping moves.

Therefore, the goal cannot be achieved, since these two facts contradict each other.

Guess the Phone Number

phone: 406.729.4381

The missing digit is 4. All the digits in the number go along chess knight's moves from key to key of the phone.

Magnetic Tetrahedron

Each face of the tetrahedron holds four coins, so you can affix 16 coins to the magnetic tetrahedron, as shown.

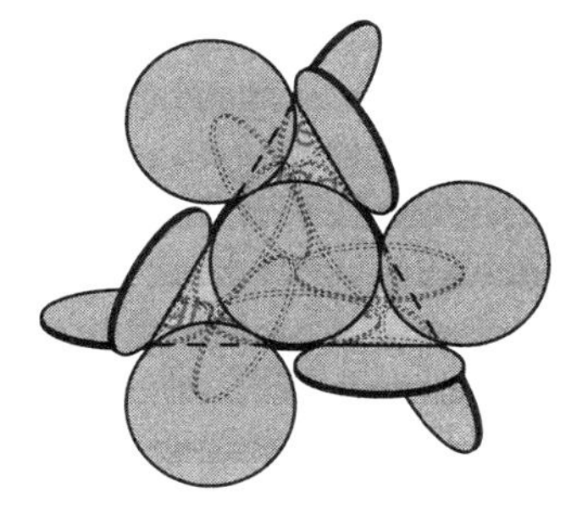

Tetrapaving

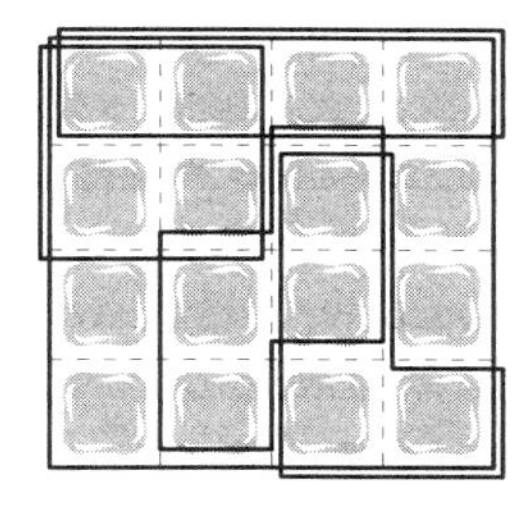

The Yawl

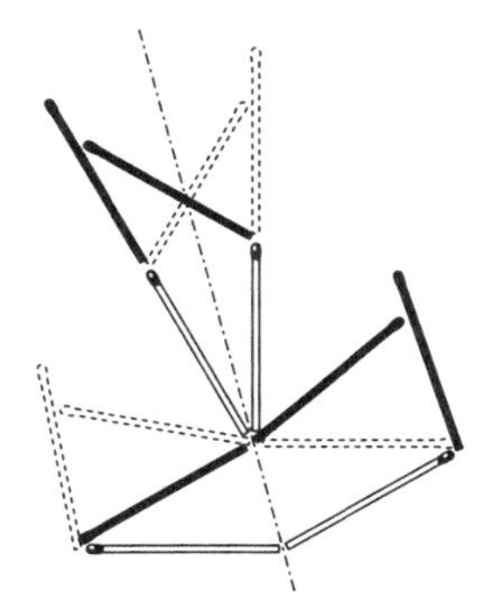

Stairs in the Pyramid

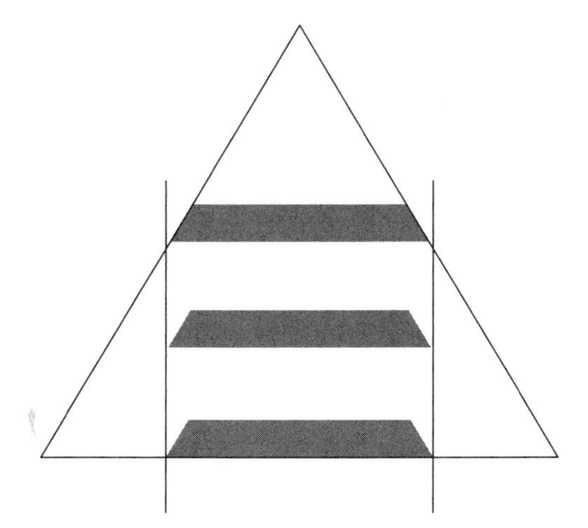

The bottom step is the widest.

Cubius

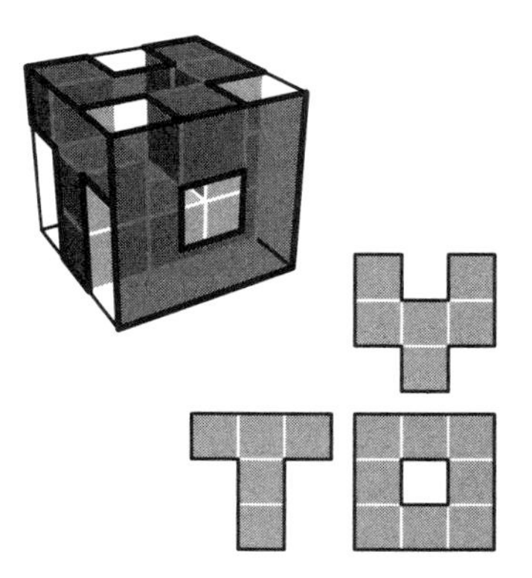

Looking at the cube from three different directions, you can see three letters which spell the word TOY.

Penta Duo

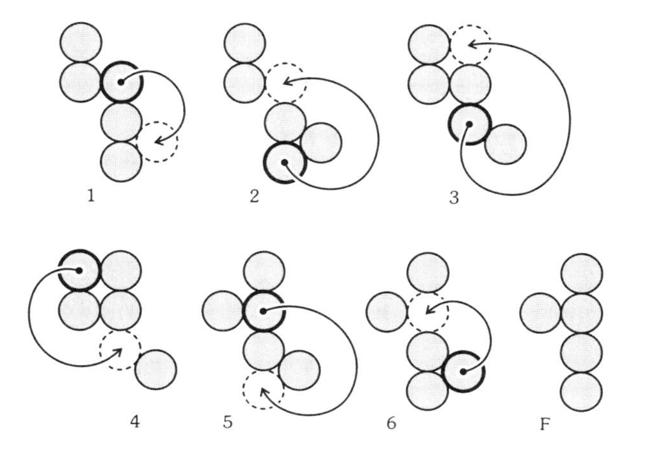

Solution to main puzzle: 6 single-moves

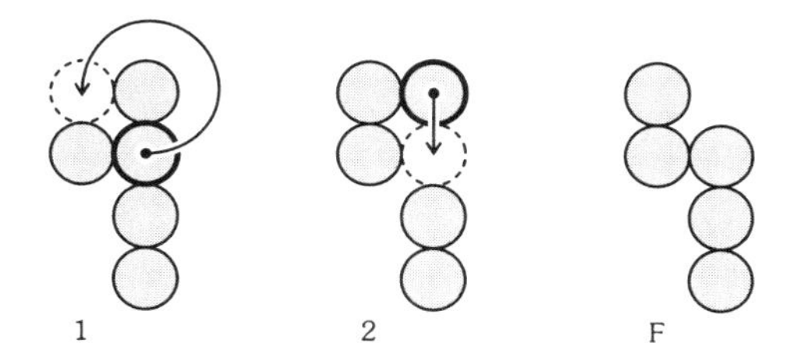

Solution to backward puzzle: 2 single-moves

The Jigsaw Square Fusion

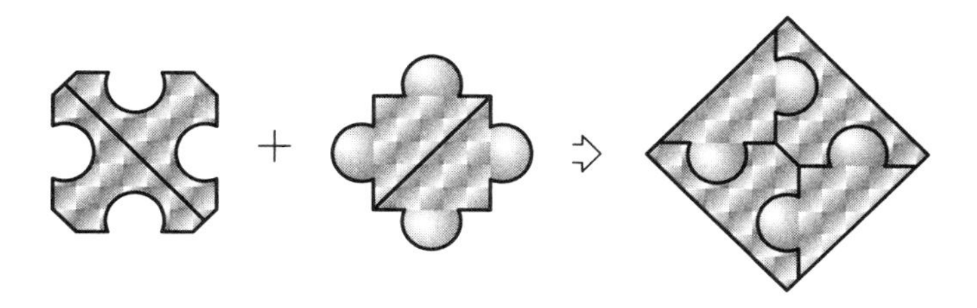

Brick Knights' Swap

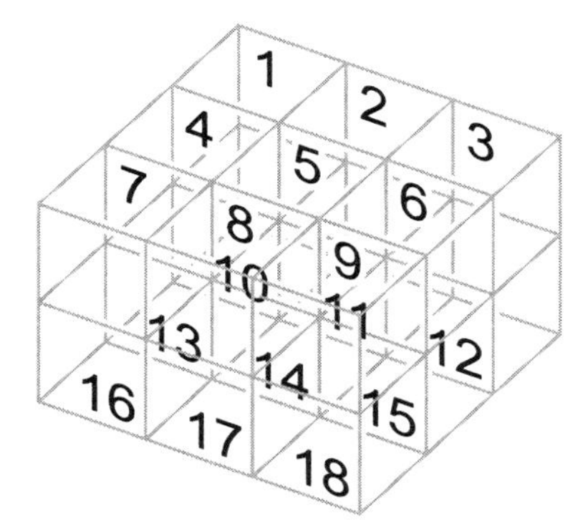

To exchange their positions, knights should perform nine moves as follows: 18-13, 7-18, 13-6-7, 15-16, 12-9, 4-15, 1-12, 16-1, 9-4.

Solid Chain

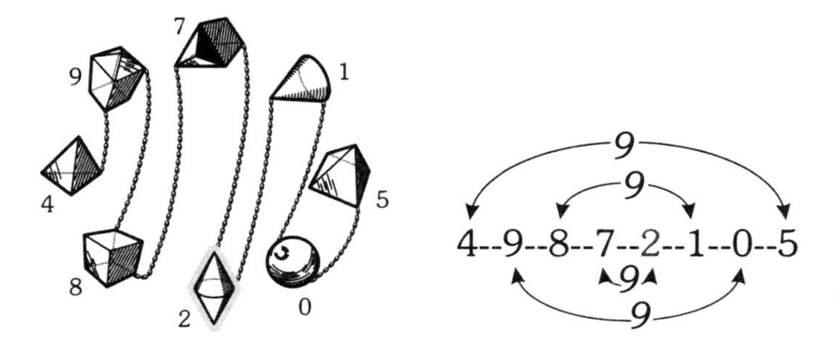

Each shape in the chain has the number of vertices indicated by the number next to it in the illustration on the left. When the chain is completed with the missing shape and then unfolded, as shown on the right, each indicated pair of shapes has nine vertices total. So, in order to complete the chain observing this rule, the missing solid should be double cone B, since it is the only solid with two vertices.

The Tomahawk

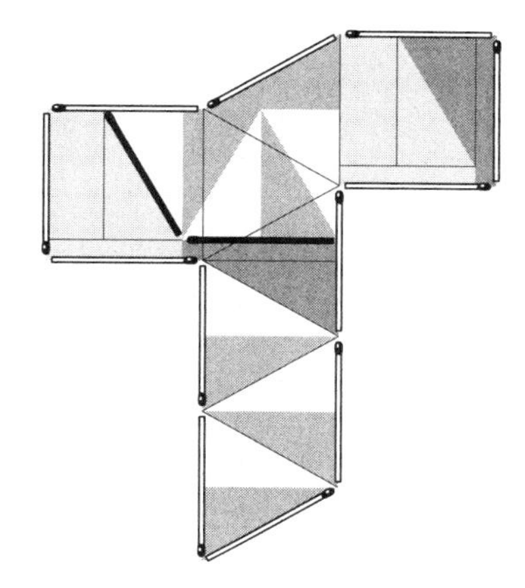

Delta Cube Score

On the surface of the cube, there are 36 triangles of three different sizes which are shown in the diagrams.

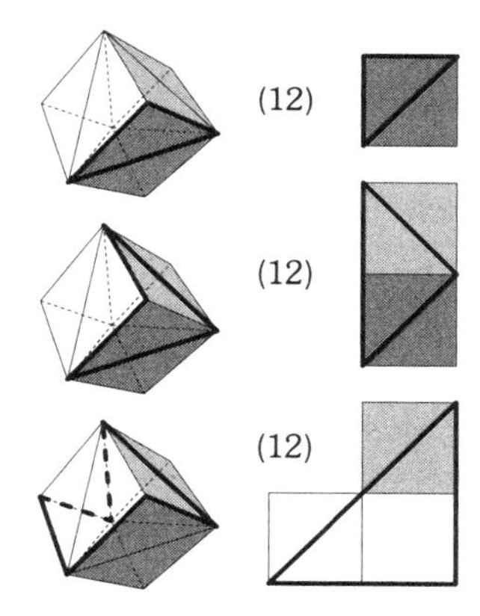

Note that the triangles have different orientations on the cube. The numbers in parentheses show the total numbers of copies of each particular triangle.

Around the Table

The family sitting around the table is the Year, and its four members are the four Seasons. Each Season's three-letter name is actually the first letters of the three months of that Season as follows:

- Spring: (M)arch, (A)pril, (M)ay;
- Summer: (J)une, (J)uly, (A)ugust;
- Fall: (S)eptember, (O)ctober, (N)ovember;
- Winter: (D)ecember, (J)anuary, (F)ebruary.

The months are arranged counterclockwise starting from December in the lower-right corner.

The Three I's

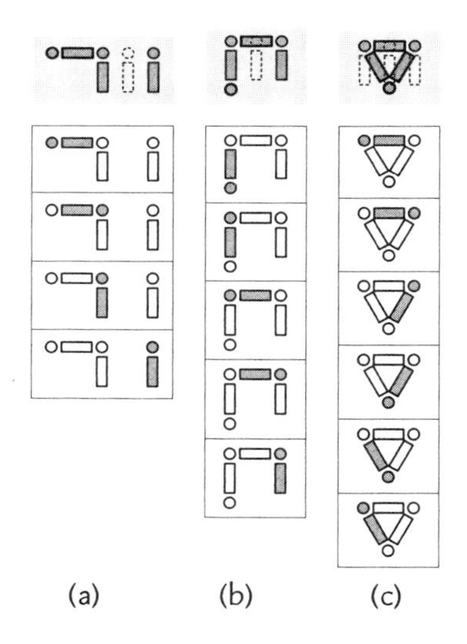

(a) Four i's are formed by moving two pieces.
(b) Five i's are formed by also moving two pieces.
(c) Six i's are formed by rearranging four pieces.

Butterfly Differing

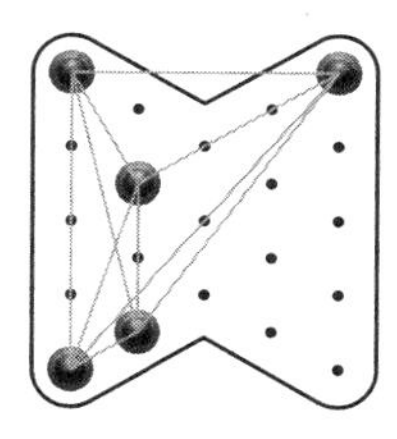

NumCount

The digit 5 should replace the question mark. When the entire expression is turned upside down and all of its digits are read like letters, but backward (from right to left), then the following sentence can be read: “She holes his shoes.”

Four Matches & Nautilus

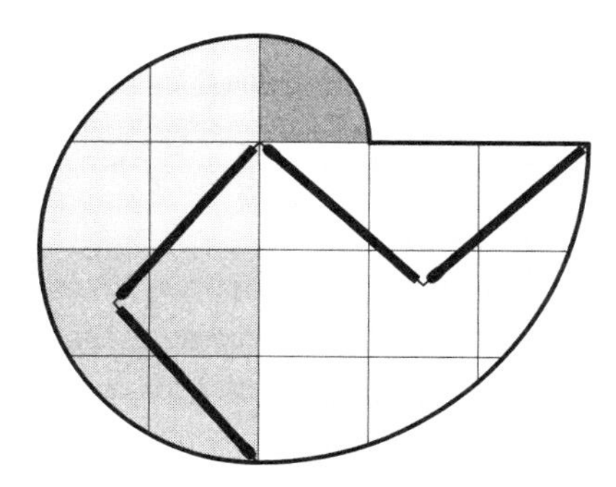

Mag³netic

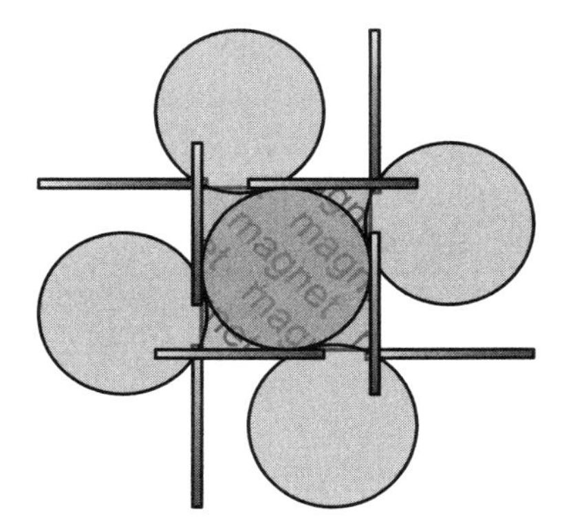

Each face of the cube holds five coins, so you can affix 30 coins to the magnetic cube, as shown.

Coin Upside-Down 2

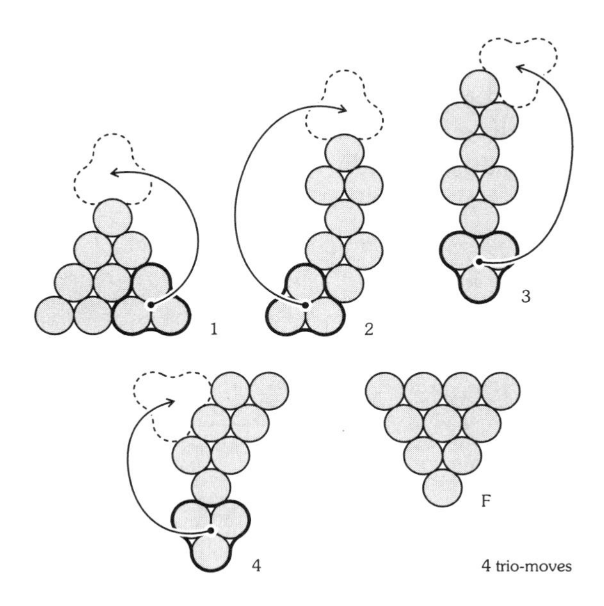

The Blue Tetrahedron Puzzle

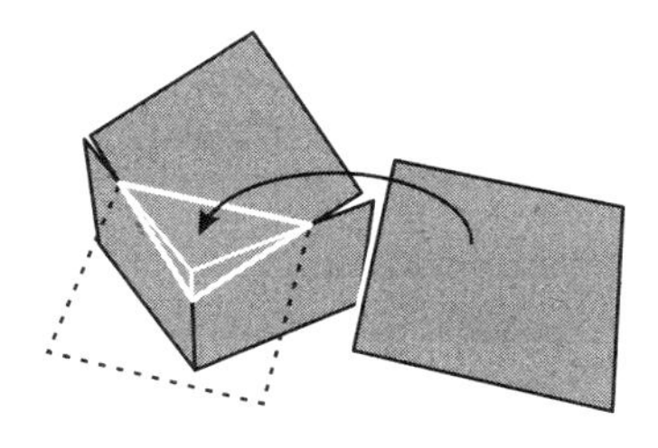

Where Is the Solution?

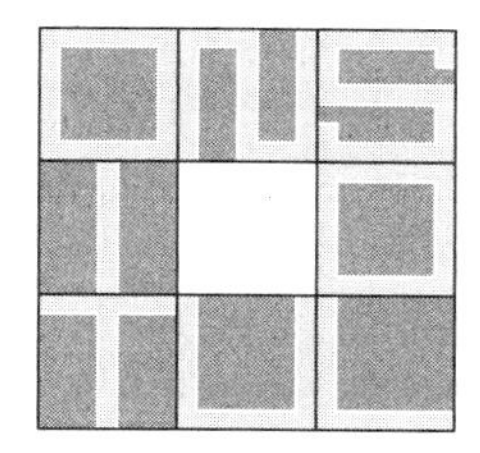

Christmas Tree

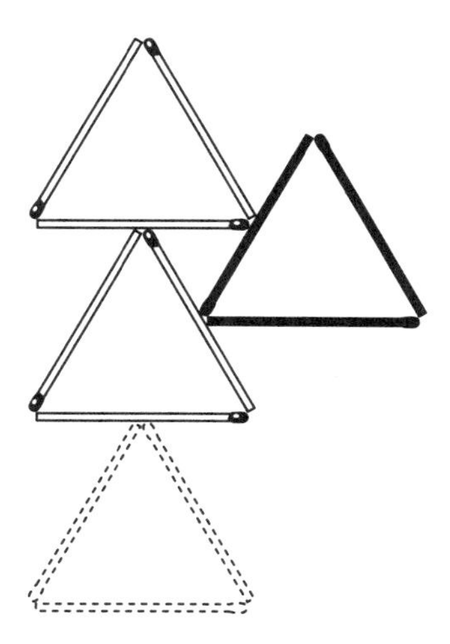

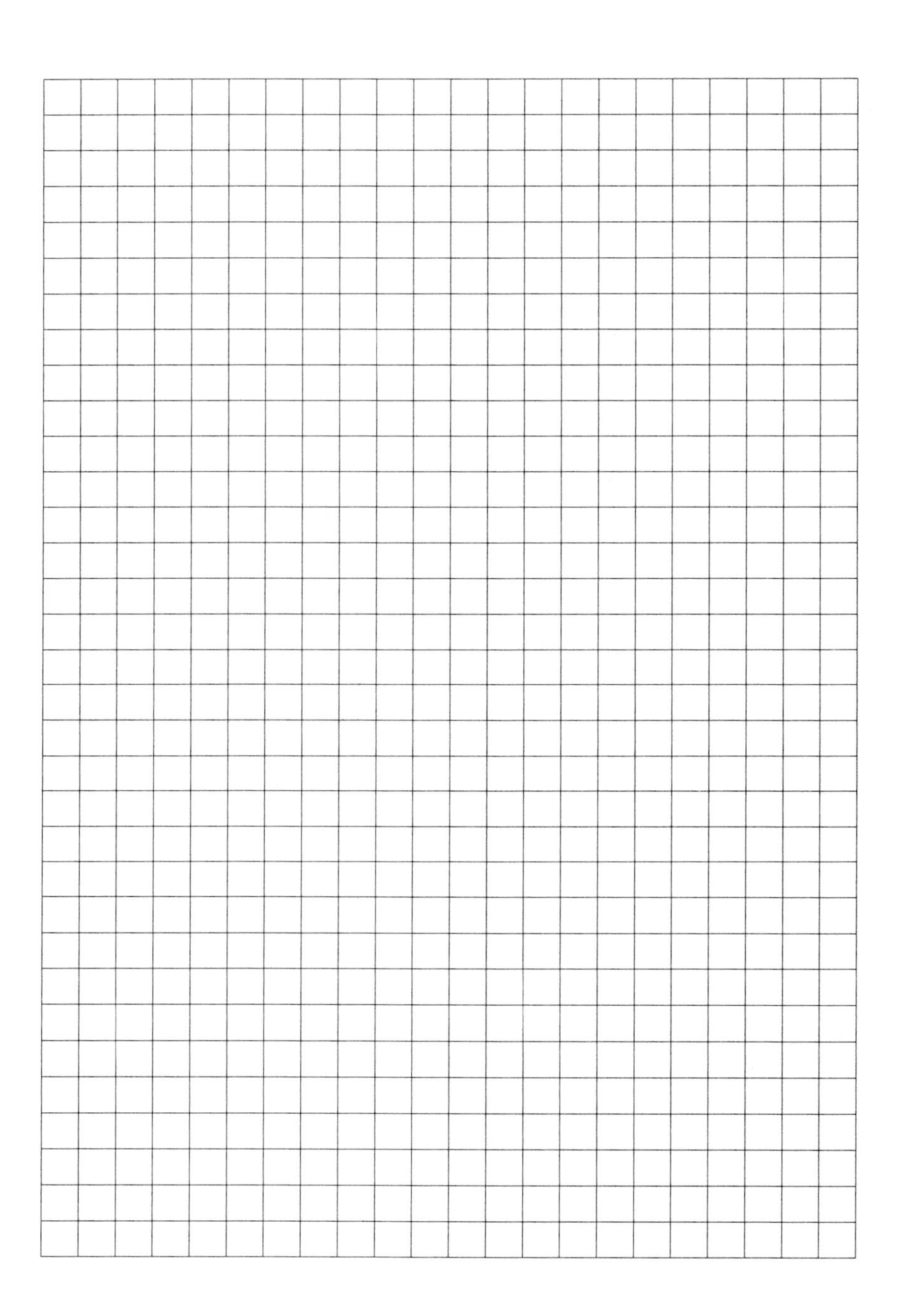